Fossils as information

*new recording and stratal
correlation techniques*

NORMAN F. HUGHES

Department of Earth Sciences, Cambridge

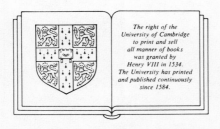

The right of the
University of Cambridge
to print and sell
all manner of books
was granted by
Henry VIII in 1534.
The University has printed
and published continuously
since 1584.

CAMBRIDGE UNIVERSITY PRESS

Cambridge

New York   Port Chester   Melbourne   Sydney

Published by the Press Syndicate of the University of Cambridge
The Pitt Building, Trumpington Street, Cambridge CB2 1RP
40 West 20th Street, New York NY 10011, USA
10 Stamford Road, Oakleigh, Melbourne 3166, Australia

First published 1989

Printed in Great Britain by Redwood Burn Limited, Trowbridge, Wiltshire

British Library cataloguing in publication data
Hughes, N.F. (Norman Francis), *1918–*
Fossils as information: new recording and
stratal correlation techniques
1. Stratigraphy. Correlation analysis.
Quantitative methods
I. Title
551.7′001′8

Library of Congress cataloguing in publication data
Hughes, Norman F. (Norman Francis)
Fossils as information: new recording and stratal correlation
techniques / Norman F. Hughes.
    p.   cm.
Bibliography; P.
Includes index.
ISBN 0 521 36656 9
1. Paleontology – Classification. 2. Paleontology, Stratigraphic.
I. Title.
QE721.′H84 1989 89–475 CIP
560′.12 – dc19

ISBN 0 521 36656 9

RB

CONTENTS

iv   *Contents*

PREFACE

As a student I was greatly impressed, as were many others, by the exploration achievements of paleontology, of which there were many distinguished exponents at the time. On entering the subject in 1947 it seemed logical to try to bring the obvious abundance and increasing availability of microfossils to bear on the exploitation problem of obtaining better correlation results in stratigraphy and in other applications. Light microscopy was then approaching peak performance and, among microfossils, plant remains such as cuticles, megaspores and palynomorphs were rapidly becoming available in great quantity. It appeared that the very large number of characters of fossils which could be discriminated and which could usually also be measured should be convertible into new stratigraphic correlation potential. The methods of taxonomy of fossils and of stratigraphic description and comparison had been in satisfactory use with all the well-known megafossils for many decades. It appeared that the degree of further success should depend on the work input and on skill in selecting new groups for attention.

The outcome of much honest effort to this end was surprisingly disappointing both from my own data and from that published by colleagues and contemporaries. Progress towards sharper discrimination appeared to be only slight. General discussion of such problems in the late 1950s and early 1960s was tinged with pessimism; 'evolution was too slow', 'fossils could not lead to any appreciably better stratigraphic resolution than that already achieved', 'other methods might well be better'. Perhaps a concentration of effort on paleoenvironments or on evolutionary patterns would be more rewarding.

Then, in the 1960s, many improved methods including electron microscopy became available for examining fossils; and data-handling was clearly due for far-reaching changes and increases of scope. This last at least removed the possibility that the cause of failure was simply the limits of human brain capacity. But again there was disappointment. Progress in the 1970s did not reflect all these considerable improvements and in many cases still does not do so, to the disadvantage of the whole subject. It is even now argued in some quarters that increased stratigraphic resolution is not only unlikely to be achieved but is also not actually needed!

Without at first fully realising that I was taking on the whole of the procedural arrangements in paleontology and stratigraphy, I started in the mid-1960s to question some of the methods traditionally applied. Through many discussions, meetings and publications in the last 20 years it has emerged that assembling the logic of necessary changes was easier to arrange than was the persuasion of my colleagues as current practitioners to consider such disturbing and supposedly heretical and dangerous suggestions. Reactions of most of these colleagues have been almost entirely defensive and have varied from steady disapproval to virtually complete opposition.

Having now assembled what I believe to be a viable consistent system, which will I consider make it possible for numerical and other manipulations to be undertaken with much higher probability of success, I suggest that it may form a firm broad stepping-stone, ahead of the present submerged slippery plank, but clearly it is by no means the end of the line. I offer it particularly to readers who are young at least in heart, and whose attitude will determine the future healthy development of both paleontology and stratigraphy as they are transformed from straightforward exploration and description into truly analytical and predictive sciences of application and exploitation. I do not claim that my scheme is the only answer, but I hope it will stimulate.

I am particularly indebted to several people who knew they were helping me, W. Brian Harland and John L. Cutbill, W. R. (Bill) Riedel and David G. Smith; to many research students and associates who have suffered; and finally to many whose mild, or even not so mild, opposition has helped to sharpen both my arguments and my determination to make this attempt. I record deep appreciation of a succession of valuable geology research grants from the Natural Environment Research Council concluding with 3/4048 in the years 1981 to 1984, and particularly of a Leverhulme Emeritus Fellowship in the years 1986 to 1988.

Norman F. Hughes   *August 1988*

Part A. Problem of effectiveness

1

Introduction

1.1   *One overall purpose.* The single aim of this presentation is to make possible the integrated use of every detail of available geologic information taken from rocks in order to achieve better resolution in sequence correlation, in paleoecologic interpretation and in logging the course of evolution. At present much paleontologic and other descriptive work is entirely compartmentalised and is usually embarked upon without closely considering such application of the information.

As indicated on the List of Contents, the first part of the book is devoted to an analysis of the problem, followed by a second part of six chapters proposing a complete revision of paleontologic data-handling in preparation for more profitable computing in semi-numerical languages. The third part, starting at Chapter 9, deals with the information-handling resulting from interpretation of the data; this is designed as a preparation of non-numerical computer manipulation through artificial-intelligence languages. The final part contains summaries and is followed by a glossary. This book is not concerned with computer programs of any kind but solely with the logical arrangements for potential input.

1.2   *Terminology.* The introduction of new ideas inevitably generates some new terminology so that fresh concepts may start life without confusion. It is also sometimes necessary to discourage use of certain older terms which may prove to be poorly defined or may obstruct. To reduce difficulty caused by these introductions and changes, a comprehensive glossary of all such terms is included at the end of this book.

1.3   *The material.* The discussion is based on fossils, for which geological phenomena the details are usually the most rich and complex in characters, thus making their evolution the more detectable; the methods are intended to be equally applicable to other geological phenomena as interpreted from rock, although that intention is not developed here. The discussion is particularly concerned with microfossils because they have often hitherto been inappropriately and wastefully treated as if they were megafossils. From the descriptive and character recording point of view there is no essential difference between any two kinds of fossil, but in numbers of specimens there is great disparity. With megafossils the problem is almost always to collect enough undamaged specimens to make a useful description with a significant record of variation. With microfossils there is almost always no excuse for failing to collect enough specimens to express variation; but the resulting sampling is so restricted and local in contrast with the total numbers of specimens available, that the method of selection becomes all-important in deciding how representative the selection from a sample may be (Gluzbar 1987).

Further, each set of specimens is by custom normally consigned to one specialist 'expert' who frequently neither investigates nor allows for parallel work in adjacent fields; it is difficult therefore for individual paleontologists to plan for integration of results from the many different types of fossil.

1.4   *'Economics' of paleontologic exploratory work.* It is perhaps reasonable to attempt to establish the general pattern of purpose behind the frequently laborious handling of microfossils; it has been argued that many of the specimens now studied could as profitably stay unattended either in or on the ground until they happen to be specifically required. On the other hand, because of the very slowly acquirable expertise that is needed for study, it becomes economically sensible to build up and hold information on fossils (and on most other characters of rocks) at a general state of readiness from which rapid expansion of knowledge to meet new demands becomes feasible. Historically, a balance of 'preparedness' has been built up in the world between institutions and universities, and among amateurs, and this appears to work satisfactorily in the present environment. It seems possible, however, that with greater efficiency of data-handling, and consequently more interchangeability of knowledge among paleontologic 'experts', some economies might be possible. Amateur paleontologists, for example, could be positively encouraged to undertake the widespread production of new observation records to revised higher standards. These could then be used in much more effective exploitation work than is currently feasible.

1.5   *Obstacles of custom.* In order to make possible the simultaneous advancement in many languages and in many countries of complex studies of interpretations of fossils and of the applications of the results, some universal customs of procedure are necessary for all to follow; these customs have been elaborately developed for more than a hundred years but the problem is the subsequent difficulty of updating them, once they are in place.

The central difficulty is the current 'cluster' approach to selection and employment of material from seemingly very great variety, and the rough approximation type of result to which it leads. Species and other taxa of fossils are clusters of specimens and in general biozones are clusters of information. Parsimony in erecting taxa, known as 'lumping', is a cautious habit thought useful for assisting the human mind in attempting to encompass the scope of nature.

Secondly, most paleontologists believe very generally in 'biologic evolution' but are also for the most part, because of their individual inability to provide any convincing proof of it, uncertain how to proceed in addressing it. In seeking proof an analytical approach is necessary, and this may be correlated with discrimination and a 'splitting' approach to the erection of taxa.

There is even a philosophical argument for splitting and against lumping, regardless of subject matter. If splitting or lumping were each successfully achieved, there is no case for or against either, but if either process be flawed then: (a) the lumped descriptive data will need to be re-observed to make the necessary discriminations, and (b) the split descriptive data may be lumped in an automatic synonymy. Thus splitting is more efficient if it is not too laborious, which is unlikely when computer assistance is available with handling of the results.

Despite the availability for some time now of rapid data-handling procedures which make 'splitting' a much more feasible approach, custom has had the effect of holding many workers back in the 'cluster and lumping' age. This has maintained a pointless rather feeble continuing exploration instead of the effective development and exploitation which the present state of earth science demands.

1.6   *Scheme to fit the purpose.* The Paleontologic Data-Handling proposal made here aims to pay more attention to the making of fuller re-usable records of each individual occurrence of fossils than to the aggregation of these records by 'lumping' into taxa, which are entirely human constructions and so much more liable to be flawed. Correspondingly in stratigraphy more attention goes into recording events than into their 'lumping' into zones. In neither case, however, is there any intention to jettison the taxa or the zones which are naturally the important elements of continuity and compatibility with published work of the past. The purpose is to render taxa and zones more universally useful by making it possible to reach back behind them effectively for the basic data when a new display of such data can lead to a new result.

Some of the detailed suggestions are not new but their co-ordinated presentation is. Some details of the package may appear on their own to be trivial and less urgent for consideration, but all are necessary for consistency. The scheme is addressed in particular to newly qualified paleontologists and stratigraphers who may by using it make much greater use of their inheritance, or who may be led to make better and more far-reaching proposals with this same objective in mind. Some caution may however be necessary with more established scientists who may also happen to instruct, edit or employ, and who may need more persuasion that serious change is needed.

2

Current data-handling for fossils

2.1   *Outline of current procedures.* The first serious observer of a
        new kind of fossil names a species from the best group of
specimens that can be assembled; his purpose in so doing is to
classify natural observations sufficiently to be able to recognise and
memorise occurrence that may prove to be usefully similar, and thus
to aid the scientific community. In erecting a species an illustration,
description and measurements are given together with locality
details, and distinction is drawn between the new and any adjacent
species which happen to have been already published. A single
nomenclatural holotype is selected, and a statement on variation of
specimens is provided if possible.

Following nomenclature rules, the species is attributed to a genus,
and whenever possible is fully classified among all other organisms,
living and fossil. With most authors there is a tendency to parsimony
is erecting new species, so that attribution of the new specimens to an
existing published species is always considered first.

In nomenclature the species epithet is normally stable because of
the existence of the holotype, but the generic element of the species
name is subject to change by any subsequent author with new ideas
on classification. The generic element of the name is placed before
the specific epithet and so is automatically indexed and spoken ahead
of it; consequently retrieval of data is frequently complicated by
generic name changes.

Additional specimens from the type locality or from anywhere else
in space or time are 'identified' with the species and are usually
list-recorded under the same name. These specimens are then

usually taken by most workers into the compass of understanding of the species, which thereby becomes a growing cluster of specimens. A range of occurrence is arrived at empirically, and when stabilised is ultimately taken to be a character of the species. Uncertain attributions to the species are prefixed 'cf.', but as discussed later there can be no precision in such statements.

The imprecise nature of the descriptive extent of a species or other taxon is widely accepted as unavoidable, but is the cause of failure to achieve more effective applications of the use of species of fossils. Interpretations of more than one species in phylogenetic studies only lead further away from any easy return to the original observations.

Although constraints on the interpretation of 'species' of fossils have long been understood (see e.g. Newell 1956), very little attempt has been made by the majority of paleontologists to employ this knowledge effectively.

2.2   *International Codes.* The International Code of Botanical Nomenclature (ICBN; Voss *et al.* 1983) and the International Code of Zoological Nomenclature (ICZN; Ride *et al.* 1985) are two elegant and very carefully prepared legal documents both maintained with the prime purpose of stabilising names of taxa of living organisms in order to reduce or to avoid confusion. They work well, although they are cumbersome; they are subject to continual adjustment of minutiae, and to long lists of *nomina conservanda* for cases in which full application of the rules is deemed to be unhelpful.

2.3   *Nomenclature for fossils.* Perhaps a majority of paleontologic colleagues and many editors have faith in the belief that strict adherence to these two Nomenclature Codes amounts to an essential requirement for orderly progress in both paleontologic and stratigraphic studies. The case, however, for treating geologic material in this way is seldom, if ever, argued.

2.4   *Taxonomy for fossils.* Taxonomy, as distinct from nomenclature, is believed by many of the same people to be a matter for free expression of opinion by authors. It does not seem to be widely appreciated that although nomenclature in the form of these

Codes has little influence on extant taxonomy of extant organisms, it has the effect of governing taxonomic procedure for fossils absolutely (see Section 6.2 below). This is mainly because of the unstated but important differences of nature of the data between living and fossil material, which will be discussed in the next chapter.

2.5   *Control through the Codes.* The Botanical Code is controlled through independent nomenclature sessions at each International Botanical Congress (currently Sydney 1981 and Berlin 1987). Paleobotanists originally, up to 1959 (Montreal), kept their affairs separate from those of the workers on living plants in a distinct Code 'Appendix for Fossil Plants', which recognised organ-parataxa. However, they have since, on an egalitarian wave of fashion, allowed this useful distinction to lapse; as a consequence paleobotanists are now so heavily outnumbered at meetings concerning the Code that effective thought about separate attention to problems of fossil plants has withered.

The Zoological Code, because the appropriate International Congresses were discontinued some years ago, has had a committee (with relatively stronger paleontological representation) which has patiently given rulings on problems over the years; because with the principal exceptions of fossil vertebrates and insects, the need there for recognition of organ-parataxa was much less than with fossil plants, treatment of fossils has remained closely similar to that for living organisms.

Thus, from different circumstances and reasoning, unquestioned orthodoxy prevails in both fields and various editorial attempts have been made to enforce it as an apparent simplification of procedure. As will be seen below, such nomenclature control dictates taxonomic style which inevitably and undesirably restricts enterprise in all data-handling for fossils.

2.6   *Holotype concept for extant organisms.* Because a species of living organisms is in constant growth and minor change, making all individual organisms transitory, a single preserved nomenclatural holotype is required for reference. This specimen remains as the 'legal' point of reference for the name in resolving

disputes and uncertainties of identification, but it does not necessarily reflect the mean of variation of any character of the organism
(see e.g. Fig. 2.1). Because the definition (description) of the living
species remains verifiable by observation, such usage has proved in
the past to be adequate and satisfactory. There is, however, the
minor theoretical difficulty that the nomenclatural holotype was in
most cases collected several generations ago and in the course of
continuing evolution some change in the populations of a species will
have occurred since that time; because such change will be very small
in scope, although perhaps potentially increasing, it is by custom
ignored.

2.7   *Holotype concept for species of fossils.* It is attractive to
        copy this satisfactory holotype procedure for use with fossils
because in most cases the fossil specimens themselves are readily
preservable and durable, although they are often very far from
representative of any complete organism. There being no question
of further change to such material, stability appears to reign.

Fig. 2.1. Diagram illustrating normal variation of a species of
extant organisms, deemed to occur on a single Holocene time-
plane. The nomenclatural holotype (H) is customarily a
complete and well-developed specimen selected by eye; it is not
related to any detail of character variation, and could in theory
contribute to any part of the curve but is usually a little larger or
more prominent than the mean.

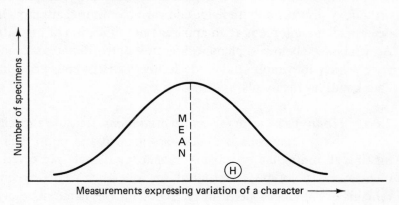

The Nomenclature Codes provide understandably (but unhelpfully to a paleontologist believing in evolution) that when fossil and living specimens appear under the same name, the holotype and name shall be taken from the living or recently dead (ICBN, Art. 58; ICZN, Art. 20, in a less comprehensive sense). There is also the problem (Fig. 2.2) with fossils representing organisms changing through time, that a single morphotype may represent different parts of the variation of successive 'species' in conformable strata, and so mislead when only small numbers of specimens are available for comparison. Such would even be the case if Philip and Watson (1987) proved to be right in their warning about the normal distribution.

2.8   *Fossil 'species' as clusters.* A species of fossils begins as a number of originally described specimens from a designated sample, from which one specimen is selected as a holotype;

Fig. 2.2. Diagram to illustrate how one kind of specimen (morphotype), as indicated here for simplicity by one character measurement (M), can be recognised at successive stratigraphic horizons although in each case it may represent a different part of the variation. This apparent constancy of occurrence could mislead in cases where numbers of specimens collected or studied had been insufficient to establish the true variation at each horizon.

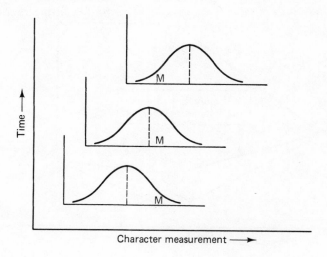

Character measurement ⟶

the precise number of described specimens affects the subsequent usefulness of a species, but not the 'species erection' process. So far, so good. From the moment of publication, it is customary for appropriate other newly discovered material to be 'identified with', or attributed to, the existing species as far as possible, because new species are in general erected with caution and even with reluctance. Any author in any country with any level of experience can make and publish such attributions to a species; there is very little development of agreed standards of recording and thus little basis for later assessment of the value of any individual attribution. In most cases the selected species name for the newly attributed material is simply

Fig. 2.3. Diagram representing the practice of species definition. The original designation of the species Xy included only those specimens within the central rectangle; these specimens (paratypes), plus others collected later from the same sample (topotypes), comprise all those which could in any sense be deemed to have come from an interbreeding unit in life. Other specimens, from different localities or from different geologic times, or both, are enclosed in a polygon and are attributed by subsequent users to the species Xy, thereby extending its meaning. Less confident attributions in the form of cf. Xy have the effect of throwing some doubt on the extent of the polygon.

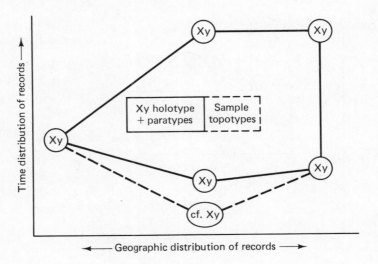

added to the floral or faunal list of occurrences for the new locality concerned (Fig. 2.3).

Custom further provides for the attributing author to express and pass on any of his doubts by if necessary prefixing the name used with the symbol cf. (confer); subsequent users of these lists then have to decide, frequently without any relevant evidence, whether 'cf.' specimens should be included positively in their interpretative operations or not. Matthews (1973) explained the earlier recommendations of Rudolf Richter, widely used in Germany and much more ambitious; although this work is addressed to the problem, it is more concerned with protecting the nomenclature system than with any use of the recorded specimens.

The consequence of such uncontrolled activities over a period of time is to alter the meaning or scope of definition of the totality of specimens included in the species under the name. The effect is inflation, and the taxon becomes an ever larger cluster with irregular limits rather than the neat concept of its generator. The name of the species usually remains misleadingly unaffected and thus apparently stable.

2.9   *Emendation of a fossil 'species' definition.* Another gradual but inevitable process of change of definition is the application of new methods of study, of which the Scanning Electron Microscope (SEM) is a current example. More accurately observed measurements and often completely newly observed characters are added; in such cases the original definition or description is formally emended under the Codes, but a notice to that effect is only appended to the name if the holotype is affected (ICBN, Art. 47) or if the name itself is concerned (ICZN, Arts. 33, 34).

Several such emendations to a single species may accumulate, but for brevity and for ease of use only the latest is normally signalled although not separately identified, and the earlier ones are subsumed. In listing attributions it is proper (Fig. 2.4A–D) to indicate which level of emendation has been followed and in theory at least this can be discovered from the dates of papers. In practice, such a procedure is space-consuming and cumbersome when applied to

Figs. 2.4A–D. A set of diagrams to illustrate changes in received understanding of the definition of a species of fossils by different authors and users over a period of several years.

Fig. 2.4A. *The traditional original description* of a species calls on paratypes and a holotype (1) from the same sample, described by an observed variation of characters (2). Other specimens (3), described from a separate sample, cannot meaningfully be included in the variation (2) unless samples (1) and (3) are deemed to be one large sample which would be unacceptable (or at least unhelpful) stratigraphically.

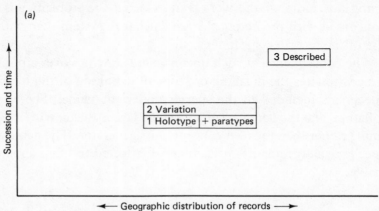

Fig. 2.4B. *Attributions by the next user.* New discoveries of specimens provide attributions (4) to a species based on variation (2) or perhaps occasionally variation (2 + 3). Doubtful attributions (5) are probably regarded as cf. (2 + 3). The dashed line envelope (---) indicates the enlarged understanding of the species definition, although theoretically the original variation (2) remains effective.

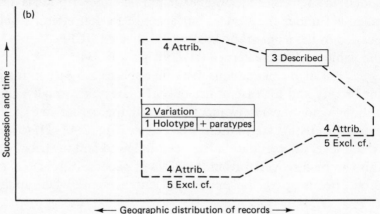

Fig. 2.4C. *Emendation by subsequent author.* This emendation arises through study of more topotype specimens from the original locality resulting in a new refined variation (6), replacing the original variation (2). New attributions (7) refer to (6), but result in exclusion of (3) and some others, thus forming a new envelope of understanding (dot-dash line, −.−.−).

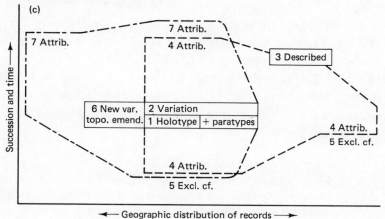

Fig. 2.4D. *A second emendation.* Further emendation arises from SEM study of topotype specimens from the original locality resulting in another refinement (8) of understanding; new attribution inclusions (9) are expressed in a new envelope (dotted line, ...). This new narrower definition results in more exclusions which affect interpretations previously accepted. It is rare for any such exclusions to be specified, so that the general tendency is for all past attributions to remain with the name (i.e. included).

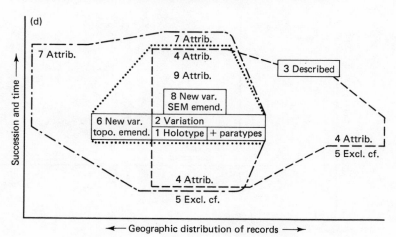

every species under discussion, and because the originally observed holotype is still in use as the sole nomenclatural reference, there is a strong tendency to quote only the original name for simplicity and to throw in all attributions (whenever made) because it is too time-consuming to discriminate; the alternative construction of a detailed synonymy is laborious and may also be discouraged in publication by editors grudging the use of space. Thus the specimen-cluster nature of a fossil species only too often becomes larger and more diffuse in character, and consequently very difficult to investigate thoroughly (see Fig. 2.4D, with attributions (3), 4, (5), 7 and 9 at different dates). The extant species is not so severely affected in this way because all attributions can at least in theory be subjected to interbreeding tests.

Fig. 2.5. *Limits of fossil 'species' in time.* The 'species' of fossils B is commonly based on a description of the character-state B as observed at its type locality. It would be more accurately represented by a described range from character-state R to character-state S, with no mention of an irrelevantly placed holotype at B. It cannot be described in terms of a difference from character-state Q to R (or from S to T) because these time-boundaries are from parent to progeny only.

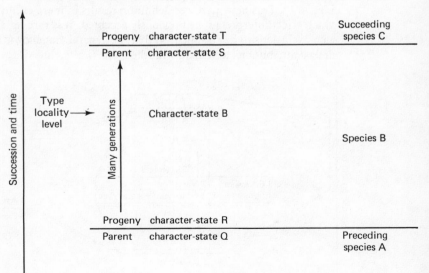

2.10 ***Fossil 'species' in time.*** Unlike extant species, all species based on fossil specimens must have limits of existence (ranges) in geologic succession (and thus in geologic time). These limits are usually arrived at empirically by procedures of attribution and comparison which do not and cannot allow for the (acknowledged) evolutionary change by natural selection within the duration of a 'range'. In Fig. 2.5, character-states Q, R, S and T may all be unavailable for observation or unobserved from collection failure, but in detail the transition Q–R must still have been from parent to progeny (and therefore a very small change; see Hughes 1978, fig. 2) and similarly with S–T. Character-state B, conventionally used to describe the species, not only lies somewhere (anywhere convenient, usually decided by availability of specimens) between R and S, but should theoretically be expanded to provide the whole range variation data from R to S (the 'range' of the species).

In practice it is only considered convenient to treat a species as an approximation for easy reference, but it is important to appreciate that in writing a description (definition) round the character-state B, character limits for R and S are automatically laid down by the nature of the species variation concept concerned and accepted; the stratigraphic range becomes a description character and thus is already decided whatever its author may believe. One of the major difficulties of paleontology is that many authors contrarily still believe that 'range of a species' is a natural attribute to be discovered only by observation; this attitude is even detectable in an advanced discussion such as that of Fordham (1986). Perhaps the main origin of such views stems from the effect of the very many gaps in all rock successions.

Another way of putting this might be to suggest that there are really two distinct species concepts, (a) a simple descriptive outline of some specimens and (b) an interpretative exercise in geologic time, and that the two are not being distinguished.

2.11 ***Cluster taxonomy.*** It has been argued that matters of taxonomy have no place in Codes of Nomenclature, and this is clearly correct. The converse is equally true, but from long development both the Botanical and the Zoological Codes now present a full mixture of taxonomic matter with the nomenclature. With living organisms the mixture appears to be adequately workable. With fossil taxa formed of ever-growing clusters of specimens lacking any potential for testing of their possible genetic cohesion, rigidity in the pattern and relationships of permitted taxa is a severe and damaging restriction.

2.12   ***The way forward.*** The best way forward appears to be to leave taxonomy to fields of activity in which it is actually necessary, to regard the Nomenclature Codes merely as an important side-issue, and to develop methods of recording through a single Paleontologic Data-Handling Code (PDHC) so that all already observed data can be retrieved in their original form and used freely in attempting to solve any stratigraphic, paleoecologic or evolutionary problem.

Part B. Proposed solution

3

Suggested new Paleontologic Data-Handling Code (PDHC)

3.1   *Purpose.* The suggestions following in this and the next five chapters are designed to provide an alternative paleontologic data-handling scheme which is complete and consistent in itself, and is thus independent of existing arrangements while remaining compatible with them. It offers the possibility of achieving much greater precision of results than has been produced before in applications such as stratigraphic correlation; these results should then fully reflect the very large amounts of detailed information available from fossils and the potential strength now provided by automatic data-processing, both coupled with a sensible view of the adequacy of the Fossil Record (Paul 1985).

3.2   *General method.* A principal aim of the new scheme has been to take the necessary discipline of record-making back to the observation stages of the field and of the laboratory, where it can be free from the minutiae and legalities of nomenclature and from the weakly defined taxonomy associated with it. At the same time every effort has been made to avoid or to reduce the need for any central registration and validating authority for any recorded observations or taxa, and to cut out the customary delays of formal publication which are inappropriate to basic observation information. In making universal observation records that all publicly bear an author's name and that are all equally treated, the responsibility on the original author for producing records of adequate quality becomes more explicit; obligatory labelling of all items including any inadequacies in basic recording means that the evidence

Fig. 3.1. Diagram showing comparison between traditional method and PDHC method of handling new material. Traditional method (above), under the label 'Identification', leads through various activities to a locality occurrence list; retrieval of information (R) is dependent on the genus 'W'. Using the new PDHC record procedure, the unique record of specimens contains all the available information and retrieval (R) may be made either through the comparison record of the taxon concerned or through the various other fields (of the GOR form: see Chapter 4). Other symbols: Ⓝ = name permanent, Ⓝ̣ = name open to change; Ⓠ = quantification obligatory, Ⓠ̣ = quantification optional.

Traditional method (identification)

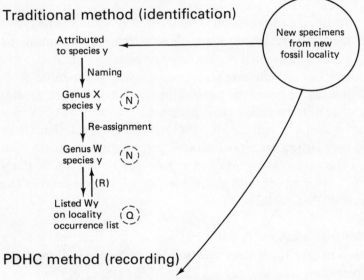

PDHC method (recording)

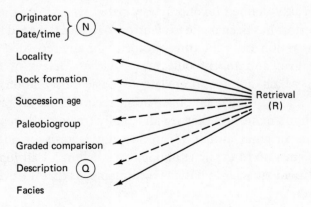

remains permanently associated with the author's name and date, and any or all of such records may subsequently if necessary be completely avoided by later workers in interpretative study in a way not possible while publication priority of names of taxa remains all-important.

3.3 **Scheme.** The Paleontologic Data-Handling Code (PDHC) is briefly illustrated in Fig. 3.1 and explained in full in Chapters 4–7. Fig. 3.2 attempts to indicate the different consequences and effects of using the existing arrangements and the proposed new Code.

Fig. 3.2. Diagram to contrast the consequences of using traditional method (left) or proposed PDHC method (right). On the left, it is customary to place all fossils in taxa in a biologic classification; on the whole this favours aggregation or 'lumping' (LL). On the right, each set of specimens is recorded separately in geologic succession; the effect is to favour 'splitting' (SS). The relative strengths of arrowed lines indicate the ease with which the interpretations indicated may be made.

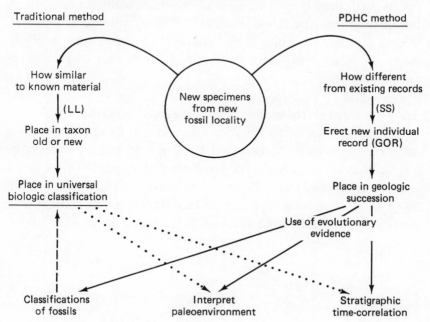

The elements of PDHC are:

Observational Records   (Chapter 4)
Nomenclature of Records (Chapter 5)
The Paleotaxon   (Chapter 6)
Replacing the genus   (Chapter 7)

3.4   ***Different approach.*** To achieve such a different approach it is first necessary to put the whole structure of Linnean taxa and nomenclature codes completely out of mind; what is proposed here is not in any sense an updating or modification of those customary arrangements. Neither hierarchy nor priority is required; latinised names are avoided simply to maintain the distinct nature of the new suggestions. A reference language is of course necessary, although only in the simplest form, but a name-form as a binominal is deliberately used to assist in retaining compatibility with current and published work. Such new language is intended to be subordinate to the handling and the use of information, and should therefore not run the risk of appearing to be its principal recognition feature.

The basis of the different system lies in the distinction of taxa of fossils, with their incompleteness as a restraint and with their distribution in time as a principal parameter, from extant taxa with natural genetically defined limits of morphology (extended even to physiology, chemistry and behaviour) but with no mention of duration of existence.

3.5   ***Some repetition.*** Because there is no acceptable logical order in which to treat the topics involved in this presentation, a little repetition may be detected both in text and in figures; no attempt has been made to remove all repetition, but the style in the sections concerned has been varied. Consequently, I suggest that if any one figure or section appears to lack clarity, it may be helpful to move on temporarily even as far as the re-statement in Chapter 16.

4

Records are primary

4.1   *The requirement.* From the point of view of any subsequent
        user of paleontologic records, the materials collected, de-
scribed and interpreted at any one locality are at least potentially
equivalent in general value to those from any other locality. Retrieval
of individual locality records, however, is seldom easily achieved
under current arrangements, or even possible (see Fig. 4.1A). The
information from each and every locality needs to be available in
complete and independent records that can be retrieved and studied
without obligate taxonomic or other reference via any other record
(see Fig. 4.1B).

4.2   *The procedure proposed.* It is first necessary to set up for all
        paleontologic material an adequate very simple procedure
(detailed in Section 4.5 below) to focus attention on the standards
required in making all primary observational records; this parallels
fairly closely the provisions of a Paleopalynologic Data-Handling
Code (PalynDHC) presented recently for a more restricted purpose
(Hughes 1986, pp. 246–248). The simplicity of this Code is such that
no permanent 'organisation', international or otherwise, would be
needed to monitor and direct it.

4.3   *The present and past custom.* In current practice specimens
        from a new locality are attributed if possible to an existing
species, usually only in the form of an occurrence mark on a species
list (see Fig. 4.1A). In most instances this procedure leaves these new
specimens without any published usable or interpretable descrip-

tion, and the custom has been reinforced by editorial policies of economy to the point at which the author may not even make such a description. In many cases the new specimens (lacking such a separate description) may be relevant to the study of a problem, while the fully described taxonomic master-material from a separate locality may not be. Such a lack of direct access to information on individual occurrences and specimens is likely when an approach can only be made through the type material of a taxon. The assembly of data from any sources for joint interpretation must result from the free choice of the scientist concerned, and should not be subject in any degree to past taxonomic decisions made by others in quite separate circumstances. It is also unfortunately true that many workers have been conditioned to acceptance of such customary difficulty and thus to the automatic restriction of the scope of their work.

4.4   *Change of emphasis.* The major change here proposed is to attach the author's name firmly to the quality of each primary (original) record of specimens from the rock, rather than only to the abstraction of a subjective collective taxon. An author may also choose to erect and name a taxon as a secondary aid to such interpretation as he happens to have in hand, but the original record always remains attributed to him and available for retrieval.

Fig. 4.1A. Diagram to illustrate current practice, in which attributed fossils are normally recorded simply as present (unquantified) or as a number (or percentage) of specimens (without further description).

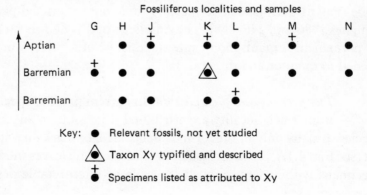

4.5   *The principles.* The principles governing the use of the
       PDHC record forms which make all worked paleontologic
data equally available for retrieval are:

(a) The record. The principal usable information about fossils
    lies in the observer's record. The record is an organised
    presentation of all information available up to the date of
    recording, concerning the specimens selected and collected.

(b) Equal importance. Each and every observation record on the
    occurrences of fossils of adequate preservation in rock has
    equal importance as a localised record of past existence of an
    organism.

(c) Numbering of records. The name of the originator plus the
    date/time provides a unique identifier; this allows the records
    to be arranged serially but requires no central registration.

(d) A rock sample is a collected unit of rock, designated and
    described; its extent is part of its specification.

(e) A sample locality is described adequately when it is possible
    from the publicly available description to locate a further and
    contiguous sample.

(f) Retrieval of data. Each record is retrievable by the name of

Fig. 4.1B. Diagram to illustrate the proposed use of PDHC
recording, in which the paleontologic records for each sample
are registered, quantified and described, and are all separately
retrievable in full.

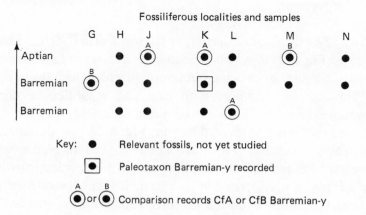

Fossiliferous localities and samples

Key: ●   Relevant fossils, not yet studied

    ⊡   Paleotaxon Barremian-y recorded

◉ or ◉   Comparison records CfA or CfB Barremian-y

the originator and the date/time of filing, by locality, by rock formation, by stratigraphic age, by paleotaxon name or equally by any one of these descriptions alone.

(g) Availability of a record. This may be effected by publication, or by deposition in a nationally recognised scientific library, or equally by its transmission to any geologist who is not and has not been a member of the institution or company to which the originator belongs. Brief notice of the existence of the record should be published in a regular scientific journal as early as possible (e.g. Publications Committee of the Palaeontological Association 1977, p. 928).

(h) Stability of description and name. A once-stored record is unalterable, even orthographically.

(i) Updating. New information of any kind, even concerning the same fossil specimens as have been used in an old record, is stored in a complete new observation record; reference to the existence of the old record is included in the new.

(j) Applicability. These principles cover all fossil material.

(k) Control. The principles above and the appropriate record forms (GOR and PTR with their instructions for use; see details below) comprise the Paleontologic Data-Handling Code (PDHC) which is designed to be suitable to be formally overseen by an organisation such as the International Paleontological Association (IPA); it is not envisaged that anything more than very occasional discussion would be necessary.

4.6   *The General Observation Record Form (GOR).* The form (Fig. 4.2) is provided as an *aide-mémoire* to an observer, as a form of storage, and as a medium through which the information can be transmitted anywhere with ease and regardless of the text language employed (because the guide letters identify the fields). The instructions for use of the form (Fig. 4.2A) are designed to be printed on the reverse (Fig. 4.2B) and include all the guidance that is necessary for such an operation. Some of the detailed instructions include points necessitated by arrangements for subsequent erection of taxa, for classification and for correlation events mentioned in the

Group	Field	ID	Content	Fig
Main identifier	Originator	A	(Name) (Address)	
	Date/time	B	(year-F) (month) (day-F) (hour)	
Rock sample	Locality	C	(Grid) (Co-ords) ±	Fig. M1
	Rock formation	D		
	Sample position	E		
	Sample lithology	F		
	Succession age	G	(Era) (Period) (Age/stage) (Chron)	Fig. M2
	Numerical age	H	Ma: ± Ma. (PMag)	
Fossils	Paleobiogroup	J		
	Graded comparison	K	Cf (Timeslot/genus) (Paleotaxon/species)	
	Number of specimens	L		Fig. M3
	Description	M		
				Fig. M4
	Variation	N		Fig. M5
	Preservation	P		
	Facies	Q		
Reference	Specimen repository	R		Fig. M6
	Earlier records	S		
	Record ends	T	(ends)	

Fig. 4.2A.

Fig. 4.2B. Entries are essential for all Records in fields A, B, C, J, K, L, R and S, but all field-identifying letters should be transmitted even when there is no entry for that field. An entry consists of the field-letter followed by all words to the right of the double vertical line, except for standard terms in parentheses (-) which may be omitted.

A Originator: *only one* author per record permitted; author's family name, followed by initials. Address of institution maintaining relevant records (author's responsibility to select).

B Date/time: date of preparation of record for use by others (*not* the date of publication). Use date/time in alternate figures and words, e.g. 1986 APRIL 12 FIFTEEN 45 only in sufficient detail to provide unique entry which (with A) acts as an automatic serial identifier.

C Locality: global latitude and longitude or (specified) national grid reference; with height in metres (or otherwise specified) with reference to appropriate Mean Sea Level (above or below).

D Rock formation: formal name (may be seismic feature subsurface).

E Sample position: quantified relation of sample position to formation boundary or to other specified level, thus indicating extent of sample.

F Sample lithology: description, normally brief if details available elsewhere.

G Succession age: formal stratigraphic scale division in descending order 'Era Period Age Chron' as far as appropriate, without terms or punctuation, e.g. Mesozoic Cretaceous Albian.

H Numerical age (including source of this): radiometric age in million years (Ma), with error range also in Ma. Paleomagnetic signature: normal, reversed or transition.

J Paleobiogroup: explicit rather than formal name for fossils or organisms, e.g. miosporestrilete, ammonoids, monograptoloids.

K Graded comparison with published taxon: CfA or CfB or CfC (no other), followed by paleotaxon name or species binominal. Note: for erection of any taxon, use separate form 'PTR'.

L Number of specimens included in the variation (N), all described from the single sample of this record.

M Description of set of specimens including quantified variation of one or more characters to indicate basis of comparison (K). Illustrations should be included on right side of form (M1 etc.), and slide/stub location data given for all microfossils.

N Variation: give details of character variation pattern among specimens.

P Preservation: include any information on re-working for microfossils.

Q Facies: include any comment on associated fossils and/or other features.

R Repository for original of this information (GOR) and for specimen material.

S Earlier records (if any): quote here originators (A) and dates (B) or relevant old record when any topotype specimen (from the original locality/sample) has been used previously in a record now updated by this new record.

T Terminates record (no other entry).

chapters following below. A worked example is provided in Appendix 1.

Maximum illustration with photographs is encouraged because the technology for transmitting these, even in colour if desired, now requires only standardisation rather than development.

4.7   *Existing records.* Many existing records can be stored on GOR forms which will clearly show any deficiencies of data. Many others, such as those on floral or faunal lists, will barely merit such attention. Records should only be transferred when required for use; mass transfer of old records to new forms is unlikely to be profitable.

5

Nomenclature/language of records

5.1   *Use of records.* Every observational record is distinct in
some degree from all others; it can only be employed usefully
in comprehension of pattern in the whole relevant field of natural
occurrences if it can be placed in a group by using similarities. The
most usual method of grouping records of fossils is 'identification'
with an already described taxon (a species) by emphasising characters in common. The problem between different observers of the
same evidence is to agree the cut-off between characters considered
to be in common and those not acceptable in that way. In current
practice this problem remains unresolved as each author (observer)
makes an 'identification' to the best of his ability and experience, and
labels his resulting attribution with the name of the species concerned. He normally contributes little or no explanation of his action
and consequently no useful assessment of it can be made by later
workers. The author may use the prefixed symbol 'cf.' to indicate his
uncertainty in attribution, but he is not obliged to explain his difficulty in any way and almost never does so.

The approximations involved in this procedure are deemed acceptable for speed in exploration work, but remain a source of
imprecision in any attempts at more accurate interpretation.

5.2   *Comparison.* To achieve any greater precision, it is necessary to limit and to date the criteria of comparison, and to rule
out subsequent alteration to any such statement. To this end a
scheme of 'Graded Comparison Records' (Fig. 5.1) has been developed (Hughes and Moody-Stuart 1967, 1969; Hughes 1976, 1986)

Fig. 5.1. Diagram to illustrate the use of graded comparison records, with material drawn for simplicity from only two rock successions. Although a conceptual envelope is drawn round the CfA records, it is only marked with a solid line where decisions have been taken (indicated by ± signs). The numbers with each record indicate a scale of desirable minimum numbers of specimens for each type of record.

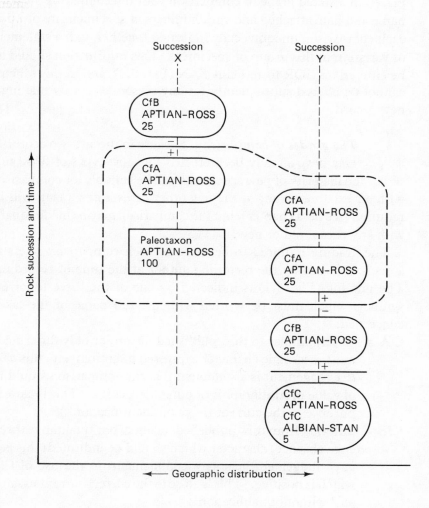

to provide a framework for authors wishing to resolve this problem or at least wishing to leave their successors with evidence on which to assess the strength of their attributions (identifications) so that these may be appropriately used or discarded.

5.3   ***The Graded Comparison Records scheme.*** The guidelines of this scheme provide that no absolute 'identification' of fossils shall be made by any author, but that all records shall be placed in a stated grade of comparison with the comparing author's name and date attached and with numbers of specimens involved in each comparison unequivocally indicated together with a statement of variation in this group of specimens. This information should all be entered on GOR form fields K–N (Fig. 4.2), and in this scheme cannot be altered subsequently for any reason except by making a new record.

5.4   ***The grades of comparison.*** Because in the last two centuries taxa have already been erected for most types of fossil material, the majority of new specimens will be suitable for comparison with an existing taxon; in the few other cases a new taxon will be required (see next chapter), but the comparison may be made equally with any taxon, old or new.

The grading required to be given to each comparison refers to a decision taken by the recording author at the time of recording. The grading (Fig. 5.1) is indicated by one of the three following symbols which must be written with the full name of the taxon concerned:

CfA = closely similar to this 'published' taxon, or only differing at most in a single character expressed quantitatively; this *positive* judgement is a statement that the comparison would be of value in stratigraphic or other correlation. (This is close in purpose to the current usage of 'identification'.)

CfB = differing from this 'published' taxon in one qualitative (presence/absence) character which should be indicated; this *negative* judgement acts as a confinement on the use of CfA which is positive. (This is close to the correct current usage of 'cf.' without qualification.)

CfC= some similarity with this 'published' taxon, but indicates a record which should be upgraded into a new taxon if a purpose and time subsequently arise. A single record may be compared CfC with more than one existing taxon. This is a 'holding' position, neither positive nor negative. (This is close to 'sp. nov.' (un-named), but avoids the erection of a taxon without any immediate purpose.)

All graded comparisons require a statement of the number of specimens observed with the variation in any relevant character together with a full citation of the relevant taxon (species or paleotaxon) with author and date, e.g.

CfA   *Clavatipollenites rotundus* Kemp 1968,
or,
CfA   RETISULC-DENTAT Hughes *et al.* 1979

5.5   ***Refinement of graded comparisons CfB.*** It might appear that elaboration of this very simple arrangement of comparisons is a useful possibility, but this is not so generally and is only appropriate and logical in the case of the grade CfB when several records of this kind have already been made.

Fig. 5.2A. Diagram to illustrate the (optional) use of graded comparison CfB in successions; the plus and minus qualifications would be used for different character variations which are relatable to superposition and to the time arrow.

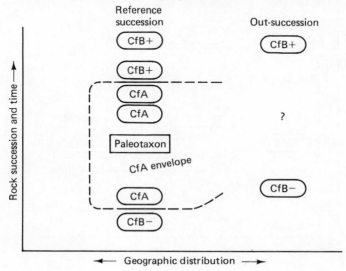

(a) In a stratigraphic context, time plus (later) or minus (earlier) may be indicated (Fig. 5.2A), if available from successional data, as CfB+ and CfB−; this may assist in such matters as lineage detection. As stated above, it is also desirable to indicate the nature of the character difference with any CfB records, e.g.

CfB+ RETISULC-DENTAT Hughes *et al.* 1979
(blunt supratectal spines)

(b) In a paleoecologic context, with records all relevant to a single paleo-reconstruction, geographic direction (NSEW) may be indicated (Fig. 5.2B) and may assist with province and other interpretations, e.g.

CfB/N RETISULC-DENTAT Hughes *et. al.* 1979 (wide murus)
It must be emphasised that no other such refinements can be made logically and therefore usefully. CfA represents a positive statement on close similarity and cannot therefore be qualified; CfC comprises only a note for possible future action. It is desirable that any development of such a CfB concept and its symbols should be kept as simple as possible.

Fig. 5.2B. Diagram to illustrate the (optional) use of CfB graded comparison on one time-plane (or approximately so) recording paleogeographic direction. Note that a CfA record cannot be qualified as it indicates positive resemblance only.

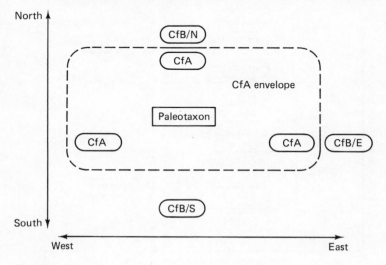

6

The Paleotaxon

6.1   *Preference for use of taxa.* An alternative to the employment
of taxa of fossils would be the use for comparison of separate
individual characters of fossils for a similar purpose (see Doyle and
Riedel 1979); although in such cases an adequate language based on
characters alone appears to be feasible, such a construction has not
been attempted here. The taxon concept for living organisms is
based on observed discontinuities in the scope of studied life. Use of
taxa has long been an accepted tradition in paleontology, and continuation of the use of taxa, although in a new form, leads to easier
compatibility with past work.

6.2   *Artificial nature of taxa of fossils.* A fundamental premise
to the use of taxa of fossils of any style is that such taxa are
recognised as being man-made (i.e. artificial). A statistical homoge-
neity of measured characters may indicate that a fossil assemblage
from a single sample approximates to a sector (population) of a
species in the sense used for living organisms, but no further or more
complete interpretation of a whole taxon is ever possible from such
single assemblages of fossils.

The limit of a taxon of fossils in time is determined automatically
by the author of the taxon through the detail of the description he
constructs (Fig. 6.1); the kind of organism concerned will have been
slowly, or even very slowly, evolving and the taxon is in effect a
time-slice (of n generations) of the relevant life continuum determined by the parameters in the description selected by the author for
the presentation of the taxon. The precise time-level of such a limit

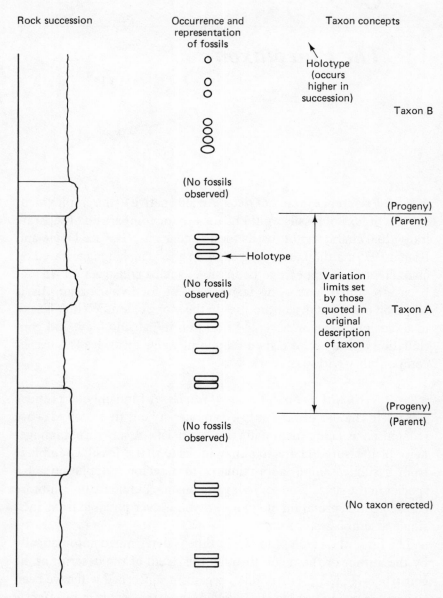

Fig. 6.1. Diagram to illustrate creation of taxa in a typically developed succession of evolving organisms, with fossils only present in suitable facies.

PDHC		PALEOTAXON RECORD FORM		PTR
Identifiers	Paleobiogroup	J		Fig. M1
	Succession age	G	(Era) (Period) (Age/stage) (Chron)	
	Originator	A	(Name) (Address)	
	Date/time	B	(year-F) (month) (day-F) (hour)	
	Paleotaxon name	K	(Timeslot) (Paleotaxon)	Fig. M2
Fossil description	Description	M		
				Fig. M3
				Fig. M4
	Variation	N		
	Number of specimens	L	Continuation sheet	
Rock sample	Locality	C	(Grid) (Co-ords) ±	Fig. M5
	Rock formation	D		
	Sample position	E		
	Sample lithology	F		
	Preservation	P		Fig. M6
Reference	Specimen repository	R		
	Earlier records	S		
	Record ends	T	(ends)	

Fig. 6.2A

between taxa may not be recorded in the field because of lack of observation of a critical part of the fossil succession, and such failure has always helped to make taxa of fossils appear to be more distinct from each other than they would be if specimens could be observed astride the limit. It also does not follow necessarily that the immediately preceding time-slice of the same life continuum must belong to another named taxon, for it may not yet have been described. This does not affect the statement that a taxon is logically limited by its description whether the author so intended and believed or not.

A further conceptual difficulty with the time-slice is that every limit of a taxon against another in time must have been drawn at a particular generation interface from parent to progeny; this means that two successive related species will not *at their interface* differ by more than the variation of parent to progeny, although the two

Fig. 6.2B. Instructions for use of Paleotaxon Record Form (PTR) for erection and unique registration of a new taxon.

The fields of description and identifying letters are the same as those used on the General Observation Record Form (GOR), but for the PTR purpose they are placed in a different order. Entries are required in all fields, except Field P which is optional. All field letters are transmitted, in the order shown on the PTR form.

J Paleobiogroup informal name (as for GOR)
G Succession age
A Name of one originator only; institutional address (as for GOR)
B Date (may be extended to time if necessary to distinguish)
K Timeslot name of stratigraphic scale division; hyphenated to uninominal paleotaxon name (e.g. Barremian-dentat)
M Description, including photographic illustrations; delete words 'continuation sheet' if not appropriate
N Variation if not included in (M)
L Number of specimens used in determining variations
C Locality
D Rock formation ⎫
E Sample position ⎬ (as for GOR)
F Sample lithology ⎭
P Preservation (if available)
R Repository for information (this record) and specimens
S Earlier records (as for GOR)
T Terminates record (no other entry)

norms or means of variation drawn for the middle of each time-slice will be distinct through being a time-slice apart in evolution.

6.3   ***Poorly defined taxa.*** An original description may have been very loosely worded, in which case it becomes inappropriate to use such a taxon for attribution or for comparison of specimens; the loose wording does not affect the facts about the time-limits, but only the precision achievable with such a blunt tool. Poorly defined taxa serve no useful purpose in this or any other field.

6.4   ***The Paleotaxon record.*** The term 'Paleotaxon' is used for a new taxon-concept developed here from the already published and used 'biorecord' of Hughes and Moody-Stuart (1969). The method of erection and of description of this new taxon is set out on the illustrated Paleotaxon regord form 'PTR' (Fig. 6.2A) and in the notes for its use (Fig. 6.2B). A worked example from Penny (1988, p. 381) is provided in Appendix 2. The form is compatible in use with the original record form 'GOR' described in Chapter 4, and indeed the paleotaxon is treated as a special case of a newly observed record. The fields of 'PTR' are placed in a different order, but are otherwise similar in scope to those of 'GOR'.

The single most significant feature of the paleotaxon is that once the record has been issued it cannot ever be emended or changed in any respect. Any necessary later comment is placed on a new PTR form (with new author/date) which is complete in itself and which incorporates a full reference to the old record.

6.5   ***The paleotaxon material.*** The record material is all taken from a single sample of any accurately stated size; all other specimens from other samples or successions are placed in separate comparison records (Section 5.3 above). The sample is designated in fields C–E on the form (PTR); its extent in the single rock succession concerned is a matter for clear decision and description. For subsequent utility in stratigraphy or in other interpretation the extent of the sample should be as narrowly drawn (restricted) as possible, but it is limited only by this consideration. The number of specimens used is required to be stated accurately, and such variation as can be so expressed should be presented with *at least* mean values and the limits observed. Taxa with high specimen numbers are desirable but more important than quantity is an honest and precise statement of account, especially necessary when numbers are unavoidably low; this accuracy will enable subsequent workers to decide either to use or to discard this taxon for their particular purpose.

6.6   ***The paleotaxon unique identifier.*** The sole (unique) identifier is provided by a combination of author's name and date in the same manner as for comparison records on GOR forms. For

Fig. 6.3A. Diagram to illustrate the development of a species into a balloon taxon. Specimens described in the species are enclosed by the dashed lines; H indicates holotype and type locality. Attributions in adjacent strata and in other strata lead for time-correlation purposes to an assumed extension of the species in time and in space, enclosed by the dotted line. Retrieval of data on records of specimens is only possible through the genus and the biologic hierarchy.

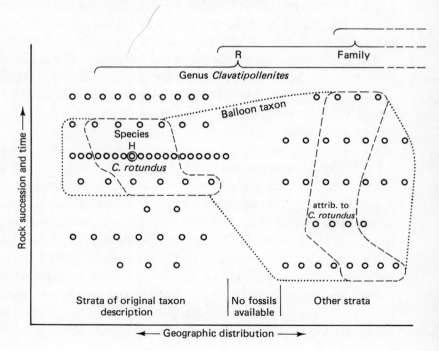

the name only one author is used (field A) because divided responsibility for such a record is a meaningless and unnecessary complication; brevity is more important. The date in field B (e.g. 1987 July 19) is expressed in a sequence only lengthened into the time in hours if more than one taxon has been erected by the author in that day. This combination is automatically unique and serial; it requires no central registration or other attention.

6.7   *The name and other identifiers for the paleotaxon.* The paleotaxon may be retrieved through its name and through its geologic succession age. The taxon name (field K) comprises a 'timeslot' name of stratigraphical scale origin (see Section 7.4 below) followed by a 'uninominal' (e.g. dentat), which is a single unlatinised pronounceable descriptive word, only long enough to be probably unique in its own paleobiogroup (field J) and thus not exceeding three syllables. The only purpose of the avoidance of latinised endings is to achieve clear distinction from (most) existing Linnean specific epithets. The word 'descriptive' in the context of this name is intended to imply a description in some sense, but not comprehensively so because this might tend to increase the length of the name. Although numerical designations for taxa have been considered for this purpose and would theoretically be adequate, a usable human language demands a pronounceable name.

The paleobiogroup (field J) is descriptive but not formal; it is intended only to limit search and retrieval, and has no rank or hierarchic position. Such an informal usage is thought to be now a viable procedure, at the present overall state of knowledge of fossils. It is not considered necessary to seek to avoid duplication of the paleotaxon name in other paleobiogroups because the group name should be quoted at least once in all deliberate reference; even accidental duplication within the paleobiogroup is not disastrous because of the unique identifier (fields A, B) on the record. Thus in this record system there is no need for special rules to avoid duplication of names throughout fossils of the plant and animal kingdoms.

The succession age (field G) may also be used as a convenient independent search/retrieval restrictor; it can be expressed down to age (stage) and chron if such control is available.

6.8   ***Stability of taxon definition.*** A classical taxon of fossils, a species, is a cluster of information to which additional material is attributed so that the scope of the cluster and thus its definition grows by accretion into a balloon taxon (Hughes 1970) of progressively less value in interpretation (Fig. 6.3A). In distinction, a paleotaxon definition is fixed and cannot be altered; all other material is placed in graded comparison records which remain dis-, tinct and can also be retrieved separately without further reference to the taxon (Fig. 6.3B).

6.9   ***Arrangements for replacement instead of emendation.*** Emendation of the definition of a classical taxon is a continuous process reflecting the progressive accumulation of more records (knowledge) and the constant improvement of techniques of investigation such as the introduction of electron microscopy and of doubtless many other techniques yet to come. Comparisons and attributions of specimens are made to the taxon at various times but the results of these activities are seldom directly relatable to each other because of the changing basis for comparison resulting from successive emendations of description.

The paleotaxon cannot be emended and consequently a new paleotaxon record replacement is made for each state of knowledge and each comparison is made with a distinctly labelled paleotaxon record.

6.10   ***Stability of the name.*** The paleotaxon name is automatically stable in that the uninominal belongs to the individual record which has been used to found the taxon and is therefore never changed. The generic element of the classical species name is subject to change resulting from any new author's opinion on classification. For the paleotaxon there is no genus; the stratigraphic scale name (timeslot) which is used instead (see Chapter 7) is stable. The only change that can be made to a stratigraphic scale name used as a timeslot is to subdivide it without losing the original, i.e. 'Mesozoic Cretaceous' can become 'Mesozoic Cretaceous Aptian' (PTR form, field G).

6.11   ***Effects of the use of paleotaxa.*** It has been suggested that
the number of taxa will greatly increase because of the replacement emendation procedure described above; in practice, however, the number of all taxa in use will not increase because many
older superseded species and paleotaxa will have been beneficially
abandoned or at least relegated to rare use for a special purpose.

Paleontologists have, over the years, become accustomed with
classical taxonomy to accepting uncertainty over the descriptive
meaning of a taxon because they have thought it unavoidable as being
a natural attribute and because they have mistakenly sought stability

Fig. 6.3B. Diagram to illustrate the use of PDHC records
to maintain the paleotaxon and all comparison records as
being permanently and individually retrievable. For
time-correlation purposes CfA records may be individually
examined, and CfB records may be considered for acceptance
or not. Retrieval (R) of data on records of specimens is
through one major informal paleobiogroup (without
hierarchy), and through the scale division 'timeslot' employed
or an adjacent one.

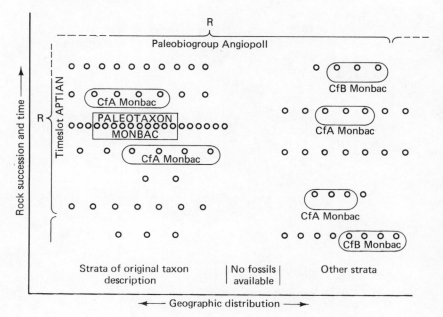

of the name alone rather than of the (more important) definition. The result of such misunderstanding has been needlessly to blur the effect of all employment of the taxon concerned in both geologic and biologic problem-solving. With the paleotaxon, that difficulty is eliminated and all subsequent use of the taxon depends on the skill and observation put into the paleotaxon description which is publicly associated with the author and date, and into subsequent comparisons, again by named authors.

7

Replacing the Genus

7.1   *The traditional genus.* The genus name, making the binominal with the specific epithet, was a valuable idea dating
back to early times and to its subsequent modification by Linnaeus;
it then obliged the original author of a taxon to make a first classification which was helpful to all users at the time of the first naming. In
the long phase of exploration of living organisms collected from the
single 'Recent' time-plane, the genus taxon has proved to be an
undoubted convenience in handling relatively similar species taxa of
unknown and increasing variety as discovery unfolded; it has at least
assisted with filing and has made possible many thought-approxi-
mations from disparate data. What is less certain and has been
curiously little discussed is whether the genus has any biologic
meaning of its own; tradition based on morphic study has provided a
comfortable belief in such a meaning, but no adequate definition of
any genetic basis has been forthcoming.

7.2   *The Genus among fossils.* Because in most cases the re-
construction of the original organism from the fossils is
incomplete and unlikely to be improved appreciably in this respect,
the genus of fossil species is a more remote concept than the genus
of the living species; any certainty of biologic meaning is most
unlikely to be provided from the genera of the paleontologic past. In
addition the (extra) time dimension complicates the position; if the
species of fossils are agreed to be entirely man-made (artificial), the
genus grouping them can scarcely possess any natural qualities
either.

Additionally, the genus of fossils as currently used (copied from the practice with living organisms) is an indefinite 'three-dimensional' cluster which cannot by its nature be used with precision.

7.3   ***The Genusbox.*** In the context of palynomorphs, which are small microfossils frequently of unknown organism-affinity and normally in great abundance, an early attempt was made (Hughes 1969) to mitigate the unhelpful continued use of certain 'long-range' genera; this was done on purely morphographic grounds through rock succession representing long periods of geologic time. With these genera of very many species comparative discussion of each already described species before the erection of a new one had become too cumbersome and unproductive to be followed; the proper handling method for new species had broken down unnoticed.

The 'Genusbox' (1969), more restricted in scope than the genus, was drawn on stated morphographic grounds and with stated geologic time-limits providing a square-box concept (hence the name) instead of the unlimited expanding cluster. This was internally consistent and successful (Hughes 1976), but its purpose was perhaps inadequately presented; it was not widely understood and it was more easily regarded as a minor heresy. It is now no longer necessary.

7.4   ***Replacement of the Genusbox by Timeslot.*** The genusbox concept mentioned above is changed here to rely only on a stratigraphic scale definition, the morphographic distinction being no longer required owing to the transfer of all priority and formality to the record alone (comparison record or paleotaxon). A new term 'Timeslot' is coined for a name (Fig. 7.1) which is now taken entirely from the formal names of the chronostratic scale divisions set out in order of decreasing rank, e.g. (Phanerozoic) Mesozoic Cretaceous Aptian, as entered on the PTR form (see field G). Thus the one word, either 'Cretaceous' or 'Aptian', would be used preceding the paleotaxon uninominal and hyphenated to it (in field K); there should be no abbreviation or coding of these words, e.g.

CfA Barremian-croton Hughes 1989 (in press)

A need to change a timeslot such as 'Cretaceous-' would only arise if the number of paleotaxa of the relevant group within the Cretaceous became too great for satisfactory search and retrieval to be possible. The timeslot name would then be changed by division to '(Cretaceous) Aptian-', and no search potential would be lost.

The use of 'timeslot' in this context does not lead to any circular

Fig. 7.1. Diagram to illustrate the use of 'Timeslot' (in place of the traditional genus) in aggregation and in nomenclature of paleotaxa. The taxon named in full as Angiopoll-Aptian-monbac may be retrieved through paleobiogroup or timeslot or originator/date. Note that specimens in comparison records may be recorded from another (usually adjacent) timeslot but the expression CfA Aptian-monbac is retained unchanged in such cases as reference to the unchangeable paleotaxon.

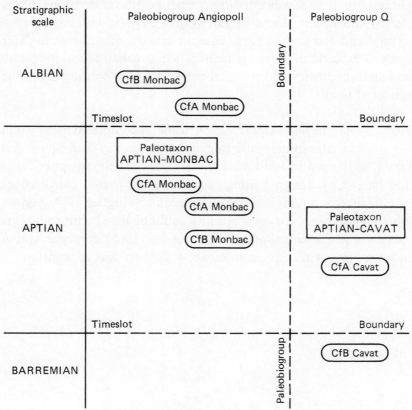

argument. Originally in exploration the recognition of stratigraphic divisions did depend on fossils, but once a complete chronostratic scale has been accepted in principle internationally (as now) the main use of fossils is in calibration and correlation, i.e. in 'exploitation'. A truly 'exploration' use of fossils in a remote situation can still if necessary begin with such timeslot terms as 'Phanerophytic' or 'Paleozoic'.

7.5   ***The Record Paleobiogroup.*** The names given here to the large and persistent groups of fossil remains are deliberately descriptive and informal, e.g. angiopoll, discoasters, monograptoloids; they are to be regarded as constituting the major heads of a general filing system, and they can be added to if necessary also without formality. Paleontologists have become so used to the apparent need to fit all fossils into higher classification of divisions, phyla, classes and orders that a hierarchy has come to be considered as natural and essential in every case; in practice attempting to determine such hierarchic arrangements is no more than a routine exercise and an unnecessary and wasteful constraint on recording the occurrences of fossils.

7.6   ***The completed Paleotaxon name.*** A completed name would be a trinominal consisting of Paleobiogroup-Timeslot-Paleotaxon, only used in full formally once in each relevant publication and having by design a filing purpose rather than a paleobiologic significance: e.g. Angiopoll-Aptian-monbac Hughes 1986 Aug 04, which if necessary for compatibility could be listed with traditional species (e.g. *Clavatipollenites rotundus* Kemp 1968) simply as Aptian-monbac, or as CfA Aptian-monbac, whichever was appropriate.

8

The Record package

8.1   *Paleontology* may be described historically in terms of three main phases, which may be distinguished as a succession, despite some overlap and despite a time-lag in some later-developed areas of the world.

(a) In the nineteenth century, universal exploration of the wonders of nature past.

(b) In the early and middle twentieth century, detailed exploration of the remarkable diversity and preservation of individual groups of organisms conducted by increasingly isolated experts within these groups.

(c) In the late twentieth century, exploitation by synthesis of accumulated information on fossils for greatly improved interpretation of succession and paleoenvironment, using automated data-handling potential for extension of scope beyond the unaided powers of the human brain.

8.2   *Traditional paleontology* of the first two phases was characterised by excitement over frontiers of knowledge and by its philosophy of the value of rapid discovery justifying various far-from-perfect means and methods. One of these methods has comprised an accidentally blinkered approach to taxonomy through nomenclature rules which has had three unfortunate consequences: (i) A continuing widespread belief that a species of fossil specimens constitutes a natural entity the true characters of which have to be sought by observation and which, once found, do not really require further description; (ii) the paradoxical assumption that because the relative ages of rocks have originally been determined from their fossil content, and because the method is often still used on new rock samplings, no taxon can be actually characterised by its known range; and (iii) a policy of parsimony in erecting taxa (known as 'lumping' as opposed to the 'iniquitous splitting'), so that human-brain data-handling should not be overwhelmed.

These three entirely avoidable difficulties, combined with a strong tendency to divide paleontology into quite separate expertise fields based on fossil groups, have weakened the subject dramatically. For example, ammonoid and dinocyst workers seldom attempt to pool their results to produce from fossils that occur together in the same rocks a single combined solution to a stratigraphic problem.

Naturally exploration continues as in phases (a) and (b) and needs full support, but its success will depend as ever on individual skills and freedom of expression; any overall control that is needed should be concentrated entirely on development as in phase (c) of finer calibration and correlation, hence the new record package offered.

8.3   *The Record package.* The Paleontological Data-Handling Code (PDHC) briefly lays down standards for production of 'Comparison Records' for all observations of fossils equally. The Paleotaxon is a precise unalterable reference taxon forming an entirely reliable basis for Graded Comparison Records. Paleobio-groups and Timeslots are necessary for orderly handling, and sep-arately for compatibility with traditional work. Priority is moved from names of taxa to descriptions of all original records and is automatically serially registered, so that no central monitoring of names or of procedures as exercised by ICBN and ICZN, nor their lists of *nomina conservanda*, are any longer necessary for fossils. Nomenclature in PDHC is treated as a distinct and necessary dis-cipline but restricted to simple allocation of names at the time of record-making on a basis of minimum elaboration consistent with filing and retrieval requirements. Even the previously mandatory publication of data is reduced to the issuing of information on the data. The question of possible individual reluctance to comply with PDHC will be covered by the discretion automatically awarded to all subsequent users to ignore records which can be seen to be inadequate for their new purposes.

8.4   *Applications of the package.* It is intended that the data stored under the PDHC rules shall be available and equally useful to any form of paleontologic exploitation work, including, if required, the assembly together of comparison records, paleotaxa and traditional species into units indistinguishable from those already employed in the past. The various new observation records would, however, also remain available in their own right for any other use independently. Published minimum standards of recording are required, not to imprison the flair of a paleontologist who has skill and insight, but to enable subsequent users to assess easily and if necessary to discard the work of others both past and present.

8.5   *Future organisation.* PDHC records are designed to be stored worldwide in the Institutions of their originators or in Regional Institutions; the GOR and PTR forms and their fields are for use in any language. Software arrangements are so far seldom compatible, and retrieval for payment across national frontiers is not yet simple, but both can be expected to improve as the need for exchange of such data increases. The present package is intended to ensure that data prepared and transmitted in this way will meet adequate standards as a scientific requirement, and are independent of the arrangements made concerning accessibility.

Part C. Applications for information-handling

9

Earth and biologic evolution

Because views about biologic evolution may enter into consideration of the topics of the next five chapters, it seems preferable to explain my own attitude briefly here rather than as digressions throughout the development of other topics.

9.1   *Explaining biologic evolution.* Even after 130 years, refer-
        ence to biologic evolution is frequently defensive and is often narrowly approached. Some biologists appear still to devote much time and even whole journals to explaining and amplifying 'evolution'. This appears to follow from excessive contemplation of genetics which is unavoidably short-time-scale based, coupled with inadequate attention to paleontology with its long-time-scale corrective; no useful connection between these two disciplines has yet been found nor seems likely soon to appear.

9.2   *Paleontologic contribution.* It is difficult for a paleontol-
        ogist to find significance in so much literature devoted to 'evolution' without, in most cases, any reference at all to fossils or to geologic history. No amount of dexterity with comparative morphology of extant organisms, even extended to cellular and biochemical matters, can successfully simulate evolutionary history. Evidence from fossils in their stratigraphic context, whatever deficiencies there may be in scope or in preservation, has to form the first main framework of study.

9.3   ***Earth evolution.*** Evolution as a general concept appears to mean 'progressive change, with the new state consequent on the nature of the old'. It is a process which applies to the whole earth and beyond, and there is no reason to regard 'biologic evolution' as anything other than a small sector of the whole. Indeed, many features observed and recorded as 'biologic evolution' are clearly an integral part of whole earth evolution, and any changes so described or discussed need to include a full account of other geologic events at or near the relevant time.

9.4   ***Biologic variation.*** Variation of characters among inter-breeding individuals of an extant species of plants or animals is in general a satisfactory common observation with adequate genetic explanation.

9.5   ***Natural Selection*** applied to this variation appears also to be an adequate explanation of all that has so far been found. In all organisms, except perhaps the Psychozoa and other higher Mammalia, natural selection amounts to the elimination at each generation of all but a very small percentage of the young organisms forming the original variation at reproduction; there is no direct competition between individuals of which so very few survive to breed. The mechanism of this 'selection' is essentially physical, instigated and controlled by geological tectonic phenomena in the broadest sense. Although in biologically crowded circumstances biological pressures may appear to be primary and certainly provide predation, it is always tectonic processes of the evolving earth which set the overall pattern of change and thus influence the composition of the successive generations of an organism which record what has happened.

9.6   ***Current evolutionary jargon.*** The word 'adaptation' is an example of totally unnecessary and even misleading jargon associated with 'evolution'. A young immature organism possesses characters resulting from genetically controlled variation; it may survive to contribute to the further gene-pool, but much more commonly does not. In all plants and lower animals, and even in mammals, survival is determined in the long run by the effects of physical earth processes, which thus provide the 'natural selection'. The characters of the organisms that survive are those that are attuned to the prevailing conditions; any organism for which this is not true is de-selected at once and thus does not survive to breed. No organism ever 'adapts' or 'is adapted', although 'man' alone may be considered by some to do so through social organisation. The continued use of this anthropocentric language for lower organisms and its further extension into 'pre-adaptation' is absurd, but the frequent use of that term illustrates the levels of fantasy which so often complicate a simple concept.

9.7   *Manipulative arrangements of 'higher' taxa.* The so-called higher taxa, such as divisions, phyla, classes, orders and families, are simply filing systems for human use. In very many cases their geologic time-limits have not been discriminated, and yet they are manipulated with zest as if their re-arrangement would somehow elucidate biologic evolution. No penetrating results are to be expected from such calculations.

9.8   *Biologic contributions.* The views expressed above are not intended as a denial of the value of the many subtle and interesting biologic contributions, but as a strong opinion that they can only succeed when blended with all available paleontologic evidence. By 'all available' is meant the inclusion of analyses of all records rather than merely of taxa and the interpretations of facies, taphonomy and preservation in every possible case. A special effort will be necessary in each instance, and no re-arranged presentation of already described fossils will suffice.

9.9   *Belief in simplicity of approach.* As a consequence of not finding, over many years, any practical use for such terms as palingenesis, neoteny, adaptation, punctuated equilibrium, clades and many more, I draw support from the views of at least some biologists (see Prance and White 1988) and I continue in a straight-forward belief that they are all artificial and unnecessary in explanation of observed biologic change as a small part of the geologist's much larger field of continuing and all-pervading *earth* evolution.

10

Proposed new Period Classification of fossils of past organisms

10.1 *Scope of a classification.* It has frequently been pointed out that a classification of any items may be successfully built to serve almost any one purpose, but that satisfactory extension to meet some further aim not originally included in the plan is always very difficult. Despite this, there remains a widespread belief among paleontologists in a possibility (just ahead) of perfecting an improved specially potent multipurpose classification for past and present organisms, which would also reflect phylogenetic history. This belief is unrealistic if it is based on any extension into the past of a classification designed for living organisms.

10.2 *Tradition and confused current practice.* Current classification of many but by no means all fossils consists of an extension into all relevant past time of a hierarchic scheme of families, orders and classes already employed with apparent success for extant organisms. With such a scheme some parts of it will be found to work at an adequate level for everyday practical purposes such as the simple filing of observations on fossils, and these successes will frequently be quoted in support of the whole classification as being suitable for phylogenetic interpretation. In sectors, however, in which numbers of species are high, the manifest overall lack of such success has led to numerous attempts at statistical improvement or more recently to cladistic manipulation. Examples of such activity following failure are in the groups of the angiosperms, the eusporangiate ferns, the bivalve molluscs and the insects, in all of which for various reasons patterns of descent are far from clear.

Failure to improve such a scheme in these ways results from attempting the logically impossible, and it is curious that comparison for this purpose has not been made with the case of human evolutionary history. Who would expect successfully to place Hittites or Sogdians, Incas or Carthaginians into a classification based on a current United Nations roll-call, with no reference at all to the passage of time?

It is instructive in this context to consider the position of the classification of the two largest groups of living organisms. The classification of angiosperm plants, with just under 300 000 extant species, remains steadily productive of strongly diverse opinions without any significant reference to fossil occurrences and without any signs of unanimity. With the Insecta, of more than 500 000 extant species, there are and have been so few paleoentomologists that practically no attempt has been made to integrate knowledge of fossils for all except a few Late Paleozoic groups, and the extant orders are usually 'traced back' without additional thought. At least with the angiosperms a struggle continues, but in the case of the Insecta complacent calm prevails.

On the other hand, in various groups of fossil animals such as the vertebrates, echinoderms, brachiopods and cephalopods there have been so many extinctions that there are numerous suprageneric taxa entirely unconnected with living organisms; in these cases a much more probably correct phylogeny is based more definitely on fossil occurrences. In the few cases of total extinction, such as tetracorals, trilobites and graptolites, such difficulties as there are seem to have been caused by misguided attempts to erect long-time-range higher taxa on narrow morphologic grounds.

One conclusion therefore is that the chief obstacle to successful advance in interpreting phylogeny is the habit of extending backward in time a well-documented suprageneric taxon of extant organisms to include fossils which at best bear only 25% of the necessary recognition characters and often strikingly less (Fig. 10.1). Advances of knowledge of extant organisms into biochemical, cellular and genetic fields lead to a great temptation to extend morphologic comparison through these additional data-sets into guesses (sometimes called 'predictions' even) of phylogeny, in which the apparent modernity of the work may appear compelling, although it usually changes nothing.

10.3   ***Proposal for a new scheme of successive classifications in time.*** In order to make it possible both to trace and to express phylogeny more clearly, it is proposed that the existing hierarchic scheme for all extant taxa should be named the Holocene 'Period classification' and should be restricted to use in that geologic epoch. Then, in each of several selected intervals in geologic time, a new independent classification should be set up to accommodate *only* those taxa and characters believed from actual fossil evidence to have existed at the time concerned; each classification would be equipped as necessary with a hierarchy of higher taxa special to that time and

Fig. 10.1. Diagram to illustrate the common anti-evolutionary practice of attributing a Cretaceous organ-fossil such as a pollen grain (Tricolpitesstriat) as a member of an extant family and consequently representative of the hierarchy concerned above the family. Clearly a pollen grain, however well preserved or well investigated, carries a very small proportion of the characters of such higher taxa.

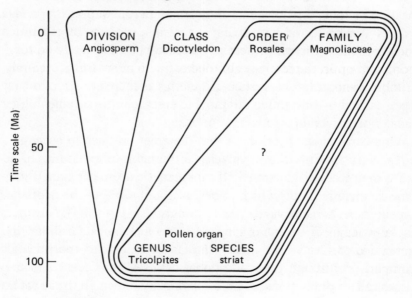

sufficient *only* to include the numbers of base-taxa actually recorded. No classification would include any item from any later period in geologic time.

As an example, initially all Cretaceous fossils would be included in a 'Period classification' based for instance on the Albian-Cenomanian time-division in the middle of that period (Fig. 10.2A,B), with parallel arrangements in the Jurassic and in the Paleogene periods.

With these new classifications, it would then be possible to reconstruct evolutionary histories of some groups of organisms by reference to a succession of purely factual classificatory arrangements which remained entirely independent of each other and of any theories about lineages or other associations. There would be no embargo on speculation or theories of any kind but they would be treated as ephemera, and the steadily improving period classifications would remain unencumbered and ready for use in the next generation of theories. For any one past time, a clear statement of facts could be readily retrieved.

Much phylogeny would remain unknown as is the true position now, but the extent of knowledge and ignorance would be more apparent.

10.4 *Spacing and nomenclature of the 'Period classifications'.*
    For best results it is clearly desirable not to erect too many such classifications until their purpose has become widely understood and appreciated. It is also necessary to allow for subsequent infill in geologic time with additional classifications as required by the quantities of available material. Consequently, a classification should be named and identified by a time-instant rather than by a time-interval or duration, to provide this flexibility.

Thus for the Cretaceous Period a classification would be based on the Albian-Cenomanian ages (stages) but named the 100 Ma Classification or more simply the '100-Classification'. Initially the next classification back in time might well be taken at 150 Ma, and all material falling between those times would be referred to the nearest classification in time; the 100 Ma Classification would then cover all material from 125 Ma to 75 Ma (assuming that there was a Paleogene classification at 60 Ma), covering most of the Cretaceous age. The round numbers further have the merit of automatic choice, so that narrow-based arguments for other dates preferred by individual paleontologic group specialists could be avoided.

10.5   ***Units of classification and their nomenclature.*** The basal units are in all cases paleotaxa or species expressed as binominals. The descriptive characters of these units are confined to characters raised directly from the fossil remains available. In paleobotany fossils are normally only single parts of plants (parataxa), such as leaves, seeds and pollen, and no attempt is made at the record level to reconstruct whole organisms: this is a less obtrusive problem in paleozoology although disarticulated bones, mollusc opercula, and jaw fossils are parallel examples. Subspecific categories are not usually recognised with fossil material, but if they are they may be treated as species.

Although in Chapter 7 above, the use for observational records of the traditional 'genus' element of the binominal was replaced by a stratigraphic divisional term used as a 'timeslot', a generic name can be freely used for paleotaxa and comparison records, along with species, in erecting period classifications. This use of the genus is secondary and subject to revision as necessary, but is quite independent of the records themselves and their retrieval as basic information.

A convention that a genus normally comprised about ten species and that the next larger unit (a family in the Holocene Epoch) normally comprised about ten genera would helpfully underline the truth that these taxa above the basic rank of species are *only* filing devices constructed for convenience of handling and hold no other meaning. However, adherence to such a suggestion is not an essential feature of such classifications. For categories above that of genus/timeslot, the new terms 'Subtroop', 'Troop' and (if necessary) 'Supratroop' have been coined as separate and distinctive units for a new purpose; in formal use where individual recognition was necessary, the prefix-labelled '100-Subtroop' and '100-Troop' are suggested (see Figs. 10.2A,B). Although such names may appear cumbersome, they would advertise their purpose and usage, would allow for additional classification, and would be compatible with and

Scale Ma	Age	Estimated no. of species	Genus or timeslot (10)	Subtroop (100)	Troop (1K)	Supratroop (10K)		
150	Tithon. (JR)	0	—	—	—	—		
100	Albian (CR)	<10K	Clavatipoll.	100-Subtroop Dicot.	100-Troop Angio.	—		
50	Eocene (PG)	<100K	Chloranthopoll.	50-Subtroop Laurac	50-Troop Dicot.	50-Supratroop Angio.	—	
0	Holocene	>400K	Ascarina GENUS	Chloranthaceae FAMILY	Magnoliales (1K) ORDER	Dicotyledones (10K) CLASS	Angiospermae (100K) DIVISIO PHYLUM	Embryobionta REGNUM

Fig. 10.2A. Example classification of some angiospermid fossils for Albian time (approx. 100 Ma); this is simplified and in outline only to illustrate levels of higher taxa required, in comparison with those for Eocene time (50 Ma) and with those used in the Holocene (0) classification. The steady radiation of these and other angiospermid fossils after Albian time is represented by greater numbers of species calling for appropriate higher taxa to file them.

distinct from Holocene (extant) families and orders which would of course retain their current meaning and as now bear no prefix.

10.6 *Construction of theories of phylogeny.* A by-product of the above proposals for a set of past classifications culminating in the employment of the traditional classification in the Holocene Epoch alone, is the necessity of producing an entirely new language and new terms for concepts in pure phylogeny. This could perfectly well be built around such words as descent, lineage, dichotomy and radiation, with the use of species and paleotaxa as the sole units; because of established custom it might be difficult to disentangle the 'genus' from these usages but some such term as a 'sector of a lineage' would be clearly preferable. This discussion is not intended to erect such a scheme, but only to demonstrate feasibility once the separation of concept has been achieved.

A further advantage could be the demise of the topic of 'macro-evolution', based on manipulation of the so-called higher taxa; these taxa have no meaning at the point of evolutionary production of a new species, and are principally of use as Holocene Epoch filing devices in that classification.

The newly proposed 'Period classifications' could be linked in a changing sequence which would unfold in time as does all else that is subject to evolution. The classification level in hierarchy attained by any distinct type of organism would in many cases probably rise in successive periods but could also fall; the lineages would become postulates which could be changed as opinion developed, rather than themselves form parts of a classification which would need formal attention if beliefs in lineages changed. Probably this would lead to the composition of new languages of information technology for this purpose, but once again the general tendency would be to separate from the wealth of accumulating observations all personal views and theories, as ephemera.

10.7 *Incorporation of PDHC records in period classifications.* Records of paleotaxa named in terms of (paleobiogroup)-timeslot-paleotaxa (e.g. Angiopoll-Barremian-croton) on a morphologic basis could be incorporated in a 100-Classification as they

Scale Ma	Age	Estimated no. of species.	Genus or timeslot	Subtroop (100)	Troop (1K)	Supratroop (10 K)	Gymnospermae (100K) DIVISIO PHYLUM	Embryobionta REGNUM
150	Tithon. (JR)	2 K	Classopollis	150-Subtroop Zonosulc.	150-Troop Gymno.	–		
100	Albian (CR)	1 K	Classopollis	100-Subtroop Zonosulc.	100-Troop Gymno.	–		
50	Eocene (PG)	1 K						
0	Holocene	1 K	Pinus GENUS	Abietaceae FAMILY	Coniferales (1K) ORDER	(10K) CLASS	Gymnospermae (100K) DIVISIO PHYLUM	Embryobionta REGNUM

Hughes 10.2B

Fig. 10.2B. Further example outline classification of gymnospermid fossils at Tithonian time (150 Ma) and Albian time (100 Ma) for comparison with the Holocene (0) classification. It will be seen that there is no reason for the taxon Gymnospermae to bear any higher rank than 'order' in Holocene time.

stand but could also be included in 100-Subtroops and 100-Troops alongside traditional species within genera, as 'Barremian-croton' if that proved helpful; this would not affect their status as observational records. Comparison records of grade CfA, with either a paleotaxon or a species can be treated in the same manner, but not grades CfB or CfC which in this context remain as records only.

10.8   *Conclusion.* The proposal for successive period classifications as a method of separating theories of evolutionary phylogeny from dependence on the Holocene classification of organisms does not depend in any way on the earlier proposal for PDHC observational records. The records, as shown in Section 10.7 above, are compatible with such classifications and would provide good quality data, but are otherwise independent. The classifications of fossil plants labelled 50 (Paleogene), 100 (Cretaceous) and 150 (Late Jurassic) would all be incomplete and at first sight of poor quality, but they would much better represent the state of knowledge in each case. Classifications of other groups of fossils need not necessarily be made at the same time or for the same periods, and need not refer to the same selection of geologic intervals as long as any higher taxa involved are clearly prefixed (as indicated in Section 10.5 above).

11

Paleoenvironment investigation

11.1   *Uniformity and evolution.* Understandably, in Quaternary studies the paleoenvironment conditions are interpreted from consideration of fossils deemed to be equivalent to their living counterparts; indeed the taxa employed in data-handling are based on reference types collected from the Holocene (<10ka) flora and fauna. The geologic time represented is short and such evolution as will have occurred in Quaternary (Pleistocene + Holocene; <1.6 Ma) time is regarded as too slight to be significant and is usually ignored. Almost certainly this does not affect the results desired and achieved, *within* the limits expected at present for such investigations. However, the small dangers involved in ignoring evolution on this scale naturally become more obvious when these methods are extended back into the Pliocene and by some authors through the whole duration of Cenozoic time.

In practice, the timing of the cut-off of methods between normal Phanerozoic paleontology and the Quaternary has to be arbitrary; it can only be logically selected by relating it to the precision and detail of results expected from the investigation in hand. Clearly, advances of knowledge and technique should ultimately progressively confine such uniformitarian methods within the Quaternary. Until that time, it is important that Miocene and Pliocene investigators should state clearly their aims and their position on this matter; they should refrain from any implication of 'absence of evolution' during investigations necessarily based on methods that assume no evolutionary change.

In the later parts of Cretaceous time or in the Paleogene Period, hard-pressed interpreters of paleoenvironment are often sorely tempted to use for example a pollen grain somewhat resembling (at CfB level) that of living *Sonneratia* to postulate the existence of mangrove swamps. It may even be claimed that interpretations based on such slight evidence at least attempt to give life to a story which could not otherwise be started. The danger, however, lies not so much in the tenuous evidence, as in the assumption that the coast margin environment of 60 million years ago would somehow have itself avoided evolving into the modern mangrove from a Cretaceous one which may not have borne any close resemblance to living mangrove conditions.

Thus the application of the 'evolution factor' is the most difficult part of the subtle process of interpretation of paleoenvironments. The selection of approximately relevant existing environments and the study of organisms considered to make and represent these is only the simple preliminary part of any such investigation.

11.2   *Discrimination and 'splitting' of taxa.* In order to record and investigate close distinction of organisms in both time and space, there is a need for taxa as narrowly drawn as is possible and profitable. The requirement is similar to that for developing fine stratigraphic correlations, although the nature of the taxon distinction may well differ. The traditional 'lumping' approach, which is usually adopted on grounds of simplicity of handling, is inevitably counter-productive as it reduces rather than increases scope of usable discrimination. Just as an open, uncontrolled cluster of attributions in a long stratigraphic succession leads to a useless balloon taxon in time, a cluster in space forms a 'paleofacies' balloon taxon which is also of minimal use in discrimination.

11.3   *Use of the PDHC scheme.* Most of the observational records to be used in paleoenvironment investigation will be graded comparison records related to paleotaxa erected separately elsewhere. As stated above in Section 5.4, graded comparison 'CfA' is the most valuable as it represents a positive author commitment to a similarity statement which is made voluntarily and which can be avoided if necessary by downgrading to the more neutral 'CfB'. This amounts to the 'defence' of precision by allowing for adjacent non-precision.

As elaborated above in Section 5.5, 'CfB' can and should when suitable be developed as a directional form of comparison.

11.4 *Advantages of this method.* One of the major advantages of using the PDHC scheme is that any number of additional paleotaxa and associated comparison records may be erected purely for the furtherance of the paleoecological investigations in hand. As they bear no priority like conventional taxa they can subsequently be completely ignored by workers with other purposes in mind, and only taken up again in suitable other investigations. Such records remain always retrievable when needed by author/date and by other identifiers.

11.5 *Evolution of paleoenvironments.* Although there is no dispute that some purely physical parameters of any environment will remain unchanged through time, most such factors were dependent on the interaction of erosion and sedimentation with the evolving hydrosphere, atmosphere and biosphere. For example, it is most likely that fossil soils from the Devonian, Carboniferous, Jurassic and Cenozoic times will show a progression of detailed developments reflecting the scope of living organisms available on land at these times, and in particular land-plant rooting systems with eventually inclusion of their symbionts. Therefore, the word 'soil' or paleosol, if not qualified, may well suggest a spurious and misleading uniformity of concept. Correspondingly in the sea such concepts as nearshore and offshore may appear simple in both geographic and sedimentologic terms, but are modified by changing biota of benthos and plankton and thus merit study for both geologic and biologic evolution.

12

General stratigraphic procedures

·

12.1   ***Purpose of discussion.*** The principal purpose of employing
        the PDHC is to make possible more accurate correlations of
rocks by applying the whole potential of 'fossil' evidence together to
such stratigraphic problems. In Chapter 14 below a new approach to
correlation is described.

12.2   ***General current malaise of stratigraphic study.*** Since the
        time of William Smith, stratigraphic correlation has been
mainly effected by comparing fossils in rock sequences, although
interpretations of sediments and their thicknesses, of seismic reflec-
tors and of many more local attributes of rocks have rightly been
attempted. However, in the span of one human generation, the
whole subject of stratigraphy, and particularly the application
of information from fossils, has now come to be regarded as a matter
of such unrewarding complexity that in some universities it is no
longer seriously and directly studied; in consequence, as justification for a low opinion of stratigraphy, it is commonly denigrated as a
mere data and memory study in contrast with the supposedly more
process-orientated disciplines. In fact it is a historical discipline
involving the interaction and application of many (process) systems,
but possibly may not be seen in this light until advances can be
announced.

No long search is necessary to indicate a basis for the pessimistic
opinion, and comment below on two well-exposed publications
(Hedberg 1976, Holland 1986) will suffice to represent the current
difficulties, although both works are very properly dedicated to

attempting to dispel some of the acknowledged confusion. The principal problems appear to be associated with (a) a great superfluity of apparently technical terms mostly with relatively weak definitions, (b) a lack of clarity of detailed purpose apparent in many major symposia and other publications, and (c) a widespread timid and perhaps self-serving reluctance to countenance change or even discussion of possible change in stratigraphic procedures.

It has even been suggested that lack of agreement internationally on the various national Codes of Stratigraphic Practice is caused by seriously different national approaches. Apart from some variation of emphasis between Western Europe, where exploration is very much confined to matters of detail, and most of the rest of the world, where this is obviously not the case, there is little significant difference of aim even in apparently unusual Codes such as that of the USSR (Zhamoida *et al.* 1977); it seems more likely that an uncritical accumulation of dubious terms and an unexplained desire to extend bureaucratic regulation have temporarily tilted development into obscurity rather than clarification. Using numerical expertise, an extensive recent application of mathematical techniques (e.g. Cubitt and Reyment 1982, Gradstein *et al.* 1985) has, if anything, appeared to worsen the situation because the basic logic and clarity of purpose of the original information were inadequate, and the cloud of mathematical juggling and justification has added to the gloom, despite possibly solving a few short-term problems.

12.3 *Development of stratigraphic procedures and purposes.*
  The current *International Stratigraphic Guide* (Hedberg 1976; reviewed by Harland 1977) was produced over many years of democratic discussion with great care and devotion by a large (IUGS) International Subcommission on Stratigraphic Classification (ISSC). Even the title of this body encouraged misguided concentration on increasing the systems of 'classification' rather than on clarification of purpose, and this can be seen from the beginning of the 'Guide' (table 1, p. 10); many of the 'technical' terms included there can be discarded with benefit, and some can be shown to be unworkable or are inappropriate for international (or for any other) regulation.

More suitable for detailed discussion is a later diagram of the interrelationship of stratigraphic procedures by Holland (1986, fig. 10, p. 15); this most courageous presentation attempts to give purpose to a web of ideas round a very dubious central concept of 'chronostratigraphy'. Holland's figure is reproduced here (Fig. 12.1A), and opposite it is a new diagram (Fig. 12.1B) showing only those terms considered actually useful and necessary for retention. The terms omitted are discussed individually in Section 12.7 below.

Fig. 12.1A. Re-drawn diagram from Holland (1986, fig. 10, p. 15), there entitled 'Stratigraphical procedures'. The appropriate somewhat polemical text in Holland's paper (pp. 13–14) may be read in support of this figure, although there is no detailed statement on the use of lines and arrows therein.

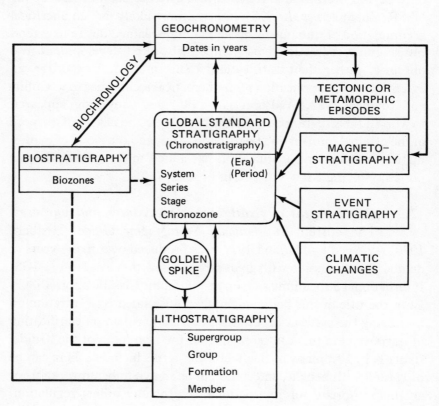

12.4 *Rock description.* In Fig. 12.2A, the primary activity is shown as organised 'Rock description' of all accessible sections of rock; the lithology, fossil content, and shape of the rock body are accompanied where available by radiometric age determination, paleomagnetic reversal observations, and any other analytical information from both field and laboratory studies. This descriptive

Fig. 12.1B. Stratigraphic procedures here expressed as comprising only four activities, after elimination of all terminology and classification no longer required.

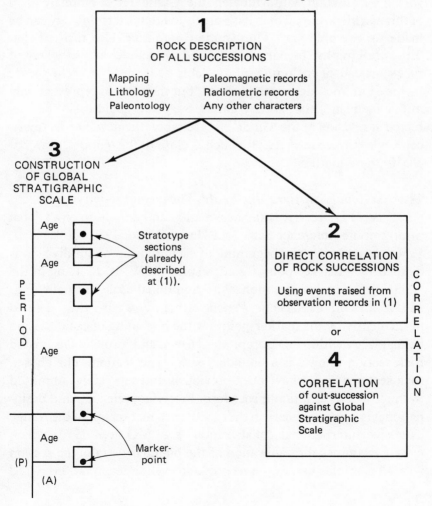

process continues with evolving techniques, and it is as necessary as it was with fossils to date the edition of the description in all sub-sequent attempts to build interpretations upon it.

As set out in the *International Stratigraphic Guide* (Hedberg 1976, p. 32), the primary unit of rock description is a 'formation' and fortunately this orderly arrangement for recording data is widely used (Fig. 12.2B) without dispute; it was derived from earlier North American Codes, and the individual definitions of the terms are both clear and subtle. Two or more formations may be aggregated into a 'group'; 'members' and 'beds' are distinctive named rock entities within a formation. Unfortunately the 'Guide' refers clumsily to all of this as a hierarchy, which it is not, although the terminology can be made to resemble one. The 'Guide Committee' also insisted that 'lithostratigraphy' be narrowly defined to include only lithology of rocks instead of including all available characters of rocks as are included in 'rock description' above, but these could represent only minor blemishes that appear unnecessary and could easily be cor-rected at revision of the 'Guide'. Regrettably, the new North Ameri-can Code (Cohee *et al.* 1978) which is close to the 'Guide', does not clarify these matters.

12.5   *Global Stratigraphic Scale.* The second activity, presented on Fig. 12.3A, is the designation and definition of a Global Stratigraphic Reference Scale (see Harland *et al.* 1982, Cowie *et al.* 1986); in this scale the beginning of each division or subdivision is defined by a marker-point ('golden spike' of W. B. R. King in the '1940s') in a rock succession. The traditional scale divisions (e.g. Mesozoic *Era*, Cretaceous *Period*, Albian *Age*) are thus defined permanently by the marker-point for the beginning of each division being placed within an appropriate selected and formally designated rock section known as a boundary-stratotype section. The chosen rock sections would have been previously and separately submitted to the 'Rock description' procedure. Preceding selection and desig-nation it is now customary in the case of each marker-point to arrange extensive international collaboration (e.g. McLaren 1977) which closely examines the calibration of the boundary-stratotype section with all possible observation records and event interpretations raised

Fig. 12.2A. Rock description. Rock successions are described independently but in terms of all available sets of characters particularly including mapping, lithology and paleontology of all kinds. The rock successions, here represented by three successions (A, B and C), are typically characterised by many irregular breaks (gaps) of non-deposition. Correlation between parts of successions is pursued opportunistically, but frequently remains incomplete.

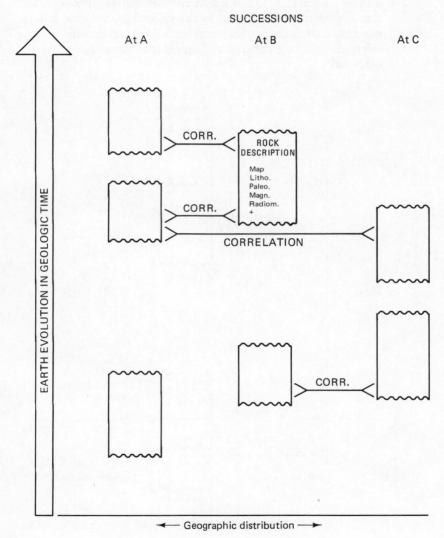

Fig. 12.2B. Organisation of units of observed and mapped rocks of a typical succession. The whole succession is compiled of *formations*, of which two are shown. Formations may be aggregated into *groups* (one shown), and these, if necessary, into *super-groups* (not shown). Within a formation, *members* may be distinguished by prominent characters, but other parts of the formation may not be distinctive and need not be named. *Beds* may be named (e.g. Q5) if sufficiently distinctive. It is expected that *formations* will be named in all explored successions, but other units are only named when needed for convenient reference; they are not required for completeness as in a hierarchy.

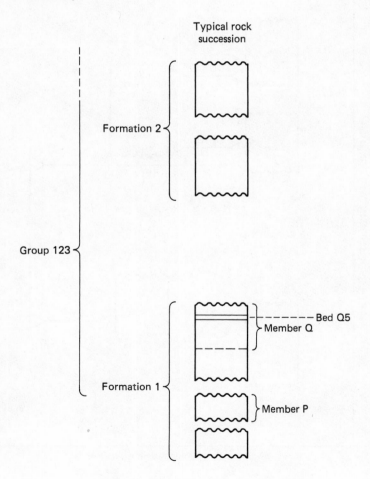

Fig. 12.3A. Diagram of a possible basis in Europe for part of the Mesozoic division of the Global Stratigraphic Scale. Marker-points placed in already described rock successions in England (E), Germany (G), France (F) and Austria (A), represent the beginnings of each period (⊚) and age/stage (●) of the scale. Each marker-point is placed, or is likely to be placed, in a selected and agreed boundary-stratotype section of rock; at some future time many more sections and points will be needed for smaller divisions of the scale. Thus far none of the points indicated has been internationally agreed, although several are regularly in use. The diagram is not intended to support any cases for selection, but is designed to illustrate the kind of pattern that is available, and some of the difficulties.

Period	Age/Stage		Selected parts of successions			
			E	G	F	A
C R E T A C E O U S	Cenomanian	Cen			●	
	Albian	Alb	●			
	Aptian	Apt		●		
	Barremian	Brm	?			
	Hauterivian	Hau			●	
	Valanginian	Vlg			●	
	Berriasian	Ber			⊚	
	Tithonian	Tth	?			
J U R A S S I C	Kimmeridgian	Kim	●			
	Oxfordian	Oxf	●			
	Callovian	Clv	●			
	Bathonian	Bth			●	
	Bajocian	Baj			●	
	Aalenian	Aal		●		
	Toarcian	Toa			●	
	Pliensbachian	Plb		●		
	Sinemurian	Sin		●		
	Hettangian	Het				⊚
T R	Rhaetian	Rht				●
	Norian	Nor				

from them. When using such sections it is important to note that the marker-point itself is neutral (Fig. 12.3B) and plays no part in correlation of other sections, but only in deciding the stratigraphic age of the result. The required formal designation is now of a Global Boundary Stratotype Section and Point (GSSP) (Cowie *et al*. 1986, p. 5).

Those division limits of the Global Stratigraphic Scale for which marker-points have not yet been agreed, are usable as an ordered sequence of names, but detailed correlation with such parts of the Scale is not possible until designation, and calibration to a satis-factory degree, have been completed. The labelled rock successions from which the Scale is derived are no different from any other rock sections in that they all need primary detailed description in the field; they are merely labelled as selected when they have been shown to contain an adequately continuous deposition of a satisfactory facies, with a suitable range of characters for correlations, through the geologic time-segment under consideration (Cowie *et al*. 1986, p. 6).

12.6   *Succession-correlation.* Succession- or time-correlation is a procedure requiring both resources and skill. The resources consist of a very large number of existing or new observation records and of interpretations from these known as events. Correlation is achieved by intersequencing the events of two rock sections or of one newly described rock section with the established global or regional stratigraphic scale. The skill, which is central to stratigraphy, has concerned holding in mind and evaluating very diverse sets of infor-mation which may provide useful events; the holding can now be automated but the more important evaluation presents greater diffi-culty as the scope of information increases with the extension of new micro-techniques and with greater concern for refinement. The customary erection of zones can be a very helpful guide in some cases, but is by no means an essential step. As indicated in Fig 12.1B, 'Rock description' and 'succession-correlation' are the only two continuous and universal stratigraphic procedures; stratigraphic scale description is an important but finite operation, although cali-bration of scale boundary-stratotypes (an application of 'Rock description') should progress with knowledge and should be updated continually as opportunity occurs.

12.7 ***Comment on superfluous stratigraphic terms.*** There is only one universal stratigraphic method as indicated in Sections 12.4 to 12.6 above; this has the one purpose of improved succession-correlation, whatever uses that may be put to subsequently. The

Fig. 12.3B. Marker-point in a boundary-stratotype section for the beginning of a stratigraphic division (e.g. the Cretaceous Period; also co-terminous with the Berriasian Age) on the Global Stratigraphic Scale. The marker-point itself is a reference with no correlation potential; all correlation is with recorded paleontologic and other events raised in the adjacent parts of the selected succession. Typically, the whole boundary-stratotype succession could lie in 10–20 metres of as nearly continuous deposition as it is possible to select. All such points are treated in the same manner, whatever the scope of the division under definition.

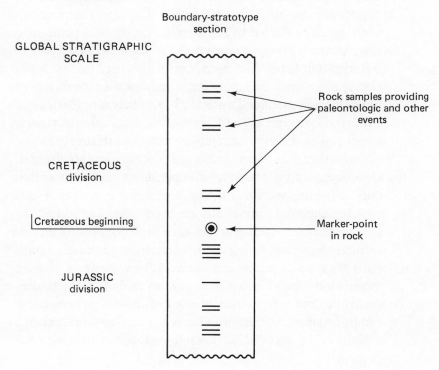

following terms used by Holland (1986) and usually also by Hedberg (1976) are either superfluous or should play a non-classificatory role simply as names for techniques.

(a) 'Chronostratigraphy'. This widely used stratigraphic term is superfluous and has been perpetuated only from excessive purposeless classification and a desire to accommodate the so-called 'time–rock' units (see Holland 1986, p. 7), a concept that has not surprisingly defied all rational attempts at usable definition. These 'time–rock units of chronostratigraphy' (system, series, stage and chronozone) have been widely used to refer to all the rocks laid down during the corresponding stratigraphic scale division (period, epoch, age and chron), but such 'time–rock' units are redundant. The confusion of a necessary standard reference scale with a traditional succession of observed units built up from the early nineteenth century is partly an accident of the subject having continued to flourish in North-West Europe where it happened also to have originated. The concentration on built-up units has led to a misguided desire to use complete type sections (body-stratotypes) for sequence-scale work, but this tendency has been neatly disposed of by Scott (1985). The more recent 'seismic sequence stratigraphy' of unconformity-bounded units (ISSC, A. Salvador Chairman, 1987) tends to raise ideas of 'time–rock' units again but this is a very general employment (approximation) related to exploration rather than to any increased precision in exploitation.

(b) Geochronometry. This term is superfluous when it describes only a technique for making radiometric time-determinations; such determinations are used either for embellishing a standard stratigraphic time-scale by one form of calibration, or for adding useful correlation potential to standard rock descriptions anywhere. When the term is used (somewhat rarely) in a wide sense to include all measurements of time such as astronomical, erosional or eustatic, it can be retained, but the problem is one of maintaining such a definition in face of a downgraded easier-to-remember version.

(c) 'Geochronology' as a separate classificatory item (Hedberg 1976; not used by Holland 1986) is a less precise name for the technique of geochronometry and so is not required. Although this term has origin in the early North American Code, perpetuation has not generated a usable definition.

(d) 'Biostratigraphy' again arose from excessive classification. Fossils, however handled, are all-important tools of correlation but they should be used, as appropriate, together with all other correlative items. There is no need for a separate 'biostratigraphy', for which an independent existence has no purpose.

(e) Biozones of various different kinds (e.g. assemblage-zones, Oppel-zones, Range-zones) are statements of knowledge and of the views of an author, complete to the time of publication; subsequently their value to others decreases through lack of certainty of definition during the inevitable accretion of further knowledge. There has also always been the uncertainty, which is very difficult to explain to those not briefed, of maintenance of the 'existence' of a biozone in face of the failure of occurrence of the zonal index fossil at any one locality. The biozone is believed by the faithful to be superior in value to other zones because of the use of the continuing change of biologic evolution, but there is no true distinction on such grounds. Consequently, the use of biozones is entirely optional and any attempt to regulate their detail is not an essential of stratigraphy.

(f) 'Biochronology' is used by Holland in his diagram but only rather oddly to label a connecting line. The use of 'biochronology' by Berggren and Van Couvering (1978) is more extensive and enthusiastic but misses the point that the contribution of fossils is to succession interpretation, which is independent of time in years. As a formal term, this is not necessary.

(g) 'Lithostratigraphy', because of the dispute referred to above about its meaning, is better dropped altogether in favour of 'Rock description'. Although its use in a broad sense to include fossil content would be acceptable, lithostratigraphy used in any narrow sense has no independent existence or purpose. Lithologic correlation is, of course, used for restricted purposes within basins and when other correlative rock characters are lacking, but it is a method of correlation not a classified discipline.

(h) 'Magnetostratigraphy'. Magnetic properties of rocks relating to magnetic poles and reversals are a most valuable and growing part of rock description. Magnetic events are valuable interpretations for calibration of the stratigraphic scales and of other successions. Because of the classification desire referred to above, 'magnetostratigraphy' was relatively recently coined but is nevertheless an unnecessary term; its purely informal use for magnetometry could stand if custom required.

(i) 'Event stratigraphy'. Events which may include interpretations raised from fossils or from tectonic, magnetic, climatic, eustatic, volcanic or other items are used widely in sequence-correlation. There is no independent stratigraphy of events and thus no need for such a separate term. Holland's (1986) use of this term appears to be peripheral to his scheme.

(j) 'Golden spike' of Holland (1986). The discussion on page 15 of Holland's paper and the use in his figure 10 show ambivalence about the purpose of a 'golden spike', and some allowance for 'isochronous surfaces' which do not belong to the practical world and would be better dropped from any discussion of methods in stratigraphic discipline. I assume, however, that Holland really uses 'golden spike' as a more glamorous name for a marker-point which is and has long been an essential detail of stratigraphic scale definition, and over that there can be no dispute. 'Marker-point' may appear more pedestrian, but the terminological glamour is not needed now that such a form of scale definition is generally accepted and now that an agreed definition of it has received IUGS authority.

12.8   ***Future usage of terms.*** Dismissal or downgrading of so
many terms as unnecessary for formal use may well appear
preposterous to those who perhaps use them regularly. I hope therefore that these terms will be allowed to disappear gradually for want
of true definition and purpose. At the same time I suggest that the
following adequately defined terms will prove to be sufficient for all
constructive purposes: Rock description and units, Global Boundary-Stratotype Sections and marker-points for the division of the
Global Stratigraphic Scale, Calibration of the stratigraphic scale,
Events, and Succession in correlation.

The changes proposed through Sections 12.3–6 and Figs. 12.2A,B
are intended to simplify the whole subject and make its use more
generally logical; then perhaps fine succession-correlation may
become more productive, and stratigraphy a more acceptable
discipline.

13

Limitations of the use of zones

13.1   *Ephemeral nature of zones.* Most stratigraphic works produce a new or reorganised zonal scheme as a mark of progress
or even as a goal achieved. What does not seem to be recognised is
that a zonal scheme only enshrines the author's knowledge up to the
time of publication, and thereafter represents a wasting asset; with
increasing other knowledge in and around the zones and without any
coherent system for their emendation, all kinds of zonal schemes
steadily pass out of date. Re-publication of zonal schemes without
updating and re-statement of all relevant knowledge, achieves nothing beyond the repetition.

13.2   *Biozones.* All zones based on records of characters changing
through successions are essentially similar in nature, but
biozones have been the most studied and more varieties of them have
been recognised. The biozones that are most precise in theory are the
phylozones (= lineage zones = consecutive-range-zones), in which
the beginning zonal boundary point depends only on *one definition* of
distinction from one species to the next in a presumed lineage. This
distinction can clearly be stated with accuracy but, as discussed
earlier (Section 6.2), a precise distinction depends on the two species
concerned having accurately delineated variations as opposed to
'cluster' definitions. Additionally it is seldom appreciated that when
fully described the point of change has to be between parent and
progeny, and consequently represents a very small step in a lineage of
descent.
    Second best are concurrent-range-zones (= overlap zones) in

which the zonal boundary point depends on the detail of the separate definitions of *two* unrelated species. Less favourable are assemblage-zones and Oppel-zones, depending on definitions of many species in each case. All other zones have even less potential for precision. Because each species definition will only stay unchanged if there is no balloon attribution of specimens to it (Fig. 13.1) and no emendation which affects it, the chances of stability of meaning for the biozone are slight when they depend on the definitions of several species. Because in the past all species taxa have normally been of a cluster nature, the biozone which uses them is in effect of a similar nature.

13.3   *Chronozone.* Hedberg (1976, p. 67) defines a chronozone as a zonal unit embracing all rocks formed anywhere during the time-range of some geologic feature or some specified interval of rock strata. This depends on imaginary isochronous surfaces and is thus unknowable as a practical unit. Hedberg (p. 68) also places the term 'chronozone' below the term 'stage' in a chronostratigraphic hier-archy, but in a footnote at once cancels this by saying that the same term can be used in a very broad and different way. Holland (1986, p. 13) amplifies the history of the use of such terms, and so emphasises

Fig. 13.1. Diagram to illustrate the unsuitability of fossil species, expanded by attributions into balloon species, for use in any fine-stratigraphic correlation.

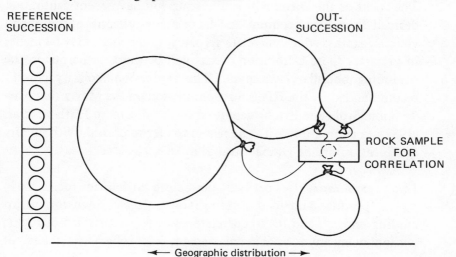

REFERENCE
SUCCESSION

OUT-
SUCCESSION

ROCK SAMPLE
FOR
CORRELATION

◄— Geographic distribution —►

the widespread confusion, even in the English language, concerning different uses for chronozones and biozones.

In Section 12.7(a) above it is suggested that this term 'chronozone' be abandoned, along with others, as unnecessary. Although this would remove the cause of confusion, the removal is not suggested solely as a quick simplification. Holland's analysis (1986) is limited although more successful, but for some reason he baulks at the final step of abandonment which would save him and everyone else so much time and trouble.

13.4   ***The Chron is not a zone in any sense.*** On the Global Strati-
graphic Scale the chron is the smallest division yet envisaged (Period–Age–Chron). Hedberg (1976, p. 69) did not support this view and dismissed the chron as simply the 'corresponding geo-chronologic term' to the chronozone; while Holland (1986), after earlier deriving the chron from the chronozone, avoided mentioning it at all.

Because a 'chron' is *not* a zone of any sort but is a small part (division) of the main successional (time) scale, its limits are defined by a beginning marker-point (as are all scale divisions) and termi-nated by the beginning point of the succeeding chron. The chron only has meaning (and existence) in the scale succession itself and has no stated or supposed 'lateral extension' to other rock successions. The rocks of the chron within the scale will have event-calibration derived from fossil content and from many other recorded obser-vations, against which correlations of any other rocks may be made; in practice, these calibration events will be most numerous in the beginning boundary-stratotype. The expression 'polarity chron', recommended by the IUGS Subcommission on the Magnetic Polar-ity Time Scale and extensively used by A. V. Cox in Harland *et al.* (1982), is compatible with the use of the term 'chron' (without any qualification) in the Global Stratigraphic Scale.

13.5   ***Summary.*** It is perfectly possible (a) to describe rocks, (b) to
produce and to use a stratigraphic scale down to and in-cluding chrons, and (c) to correlate any two successions of rocks, without using any zone or zonal concept. A biozone is a harmless but

blunt and usually unnecessary correlation tool, although in rapid exploration it has often proved to be a useful temporary concept. A chronozone serves no useful purpose at all, although the term is used mistakenly by some instead of the less familiar 'chron' for a small division of the time-scale; however, this time-scale division is in no sense 'a zone'. In exploitation work, therefore, zones are best treated as once-useful ephemera and are better avoided in any attempts to improve precision.

14

Event-Correlation

14.1   *Preference for use of events rather than zones.* In describing
       successions in order to compare them and thus to achieve
correlation, each item of description should be based on as small a
thickness of rock as possible, the single sample. This smallness
makes it at least theoretically possible to discriminate parts of successions of the same order of magnitude as these samples. When a
zone is used the information from many samples is aggregated into a
fairly large unit (the zone), below the size of which no further
discrimination is possible.

When these events are erected and recognised from single
samples, it is possible to order the events from two or more
successions in such a way as to reveal at once changes of depositional
rate or hiatuses of deposition in the successions. As all such
successions have in general more gap than deposition, it is then
possible to achieve correlation to the detail permitted by the nature
of the succession rather than by the coarseness of the zonal
scheme.

14.2   *Definition of 'event'.* The term 'event' has been used in
       stratigraphy over the last 20 years although it is not used or
even indexed by the *International Stratigraphic Guide* (Hedberg
1976); for definition an event can be said to refer to 'an interpretation
raised by an observer from a single rock sample of any stated size',
and may involve information on biological, chemical or physical
phenomena. This definition can include, if desired, the concept of an

event having long duration (e.g. a glaciation, an orogeny), although this must be with reference to the extent of the samples specified in the definition which would be correspondingly broad. Naturally events to be used in correlation will be based on a restricted sample size but it seems preferable to define the term generally and not to restrict its use in any way; within such a broad scope encouragement for the use of descriptive prefixes to the term will aid search and retrieval of published events. As an example, Hughes and Moody-Stuart (1969) used the *Cicatricosisporites* content of Early Cretaceous palynologic samples from the English Wealden for correlation purposes, and the interpretations were simply referred to as numbered events in this paper; in a comparable paper now they would be named at least as 'palyn-events' or possibly as 'Cicatricosi-events'.

As with any simple term, some uses have arisen which may appear to be out of line and thus to confuse the definition. For example, some oceanic stratigraphers have insisted that an 'event' has to be a change of state from one sample to another succeeding sample; this definition is interesting but can easily be accommodated in the definition above by including both samples together.

14.3 *Problematic use of the term 'event'.* A more awkward difficulty of definition is the common use of the incoming of 'fossil X' in two or more separate sections, which incoming is named the 'Fossil X event' or First Appearance Datum (FAD); such a use presupposes equation of the incoming of the fossil in the 'out-successions' with that in the reference succession when quite possibly, for reasons of local preservation failure or even of diachrony, this correlation may well be incorrect and certainly cannot be assumed. To maintain the definitions, the two samples in their separate successions could be brought together to be considered as one in the definition, but for many purposes this would be unhelpful because it is once more a form of ballooning of information, concealing a correlation or the need for one and thus reducing future potential for discrimination. Clearly for user advantage each event should be independent and come from a rock sample of a size well below the resolution sought in correlation.

14.4   *Independence and inequality of events.* What may appear to be a case of an event in an out-succession being equal to that in the reference succession is never a case of true equality; if events are raised from fossils at least the number of specimens seen and their associated other fossils and lithology will differ in some respects in the two successions even if great precision is attempted (Hay 1972, Fahraeus 1986, Gluzbar 1987).

In the simplest possible case of a request for stratigraphic dating of a new sample (Fig. 14.1), four alternative statements can be considered:

    (a) X is younger than R

    (b) X is older than R

    (c) X is time-equivalent of R

    (d) No decision on (a) to (c) is possible.

Of the above there is insufficient evidence to support either (a) or (b), and while (c) may appear attractive as an approximation, it is illogical because the fossil contents of X and R are not equal and could not be so. When new information is subsequently discovered at or around X or R, the position will change and the original request should be re-presented; if, however, statement (c) for equality had already been used, its effect would have been final, thereby precluding all further investigation or discussion. Only statement (d) is acceptable,

Fig. 14.1. Diagram representing the simplest possible case in meeting a request for stratigraphic dating of a new sample X in an out-succession by comparison with a described reference succession.

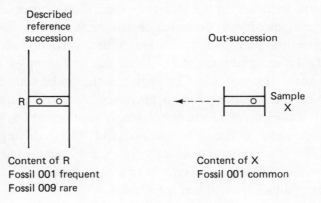

however disappointing it may appear to be. Thus every event interpreted in every succession is independent and different from all others.

14.5   ***Correlation.*** Using the background of earth evolution, including biologic evolution within it, all parts of the earth are considered to be in continuous change of both position and nature under control of gravity, isostasy, plate movements, eustasy and climate; no consistency of rate of change is, however, implied in this statement. Correlation of successions by events raised from individual rock samples then becomes a matter of ordering these events in succession and/or time.

Further, if diachrony has already been detected, it should be possible with the independent records proposed above to study it and ultimately to use the course of a diachrony positively in correlation.

14.6   ***'Bracket' correlation.*** The logical way to correlate a single event from an out-succession with a 'calibration set' of events in a reference scale succession is to attempt a pair of statements (Fig. 14.2A). Thus, taking a fuller and more representative case of a request for correlation of a new sample (X) against a reference succession (P–S), the following statements can be considered:

  (a) X is younger than P, with 95% probability
  (b) X is older than S, with 90% probability
  (c) The remote possibility of 'equality', included within the bracket formed by (a) and (b)

Although the information on sample P is quite independent and may differ entirely from that on sample S, the pair of statements ((a) and (b)) is known for convenience as a 'bracket'.

The 'bracket' can subsequently be refined (narrowed) progressively on each occasion that new information becomes available about either X or P–S (Fig. 14.2B). Correlation therefore is never complete and is constantly open to refinement as the density of sample investigation increases with the scale of observation applied. In this way finer stratigraphic resolution can be directly dependent on the investment of time and effort, and in that sense will be scale-dependent.

It can also be seen that to correlate by equating the events from two samples in separate successions is not only illogical from the outset (as stated above) because the data are never identical, but also once an 'equation' statement has been made it is incapable of any progress-ive refinement and thus remains as a frozen approximation; such equated events are often referred to as co-eval in report writing, but even that allows of no further progress. This is unfortunately the state reached in many traditional stratigraphic correlations in the past, and is responsible for much of the pessimism surrounding attempts to make finer stratigraphic resolutions. At least the nature of this difficulty was pointed out long ago by Rodgers (1959). The cluster nature of traditional taxonomy of fossils, discussed above, also works in the same sense against any progressive refinement.

14.7   *Multiple bracket correlation.* As shown in Figs. 14.3A, B, this kind of bracket correlation when applied to a sufficient number of events will ultimately reveal both stratigraphic breaks and differing rates of deposition, but only when sufficient work has been put into decreasing the size of individual brackets. It is suggested that this method will then reflect the true state of knowledge in any succession under investigation, which should be an improvement on estimates from graphic manipulations.

14.8   *Nomenclature of events.* Although it is probably unnecess-ary to subject events to the same level of description as paleotaxa and comparison records, the two principles of author/date serialisation and immutability after first use or publication are worth retaining in addition to the obvious geographic and rock formation/position data, best expressed as sample numbers. As suggested above a prefix such as dinocyst-event, ammonoid-event, tephra-event should be adequate for search/retrieval purposes with-out involving any precise classification of prefixes.

e.g. Palyn-event: Incoming CfA Barremian-monbac, War-lingham borehole, 1246 feet, Hughes 1986 October 12 (+ any necessary elaboration of the circumstances of occur-rence, such as 'negative sample at 1247 feet').

Fig. 14.2A. Diagram of a simple case of a 'bracket' correlation used to date a new sample (X). A pair of independent statements make a bracket of any size representing the degree of probability desired. Note that the whole of the reference succession between S and P is embraced, but not including the samples S and P.

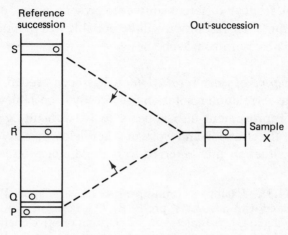

Fig. 14.2B. Diagram to illustrate refinement of the correlation bracket after receipt of new information on microfossils from additional samples A–G. At the sample spacing indicated, the correlation bracket might ultimately be reduced to R–D, but at any time it represents the current state of knowledge and the current probability requirement.

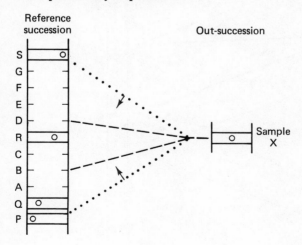

14.9   ***Types of event.*** Although most workers would by custom
operate with certain kinds of event, all events of all different
kinds can be plotted on the same sections and bracket references
need not be confined to pairs of similar event-types. Thus integration
of information from all different sources to provide a more powerful
way of obtaining finer resolution will be possible in a manner never
seriously available with zonal schemes.

14.10   ***Relevance of finer resolution.*** It has been argued that adequate correlation resolution has already been achieved, and
that as any improvement also appears to be difficult, geologists
should accept the situation and work in other fields. It is my conten-
tion that the situation thus referred to is the direct product of the

Fig. 14.3A. Examples of multiple bracket correlation. P is
correlated with post-A (Å) pre-E (E̲); Q is correlated D̊G and R
is correlated G̊J̲. Samples S and T, lacking positive characters
but usable in superposition, may be correlated initially, S to
(DE)J̲ and T to (AB)G̲.

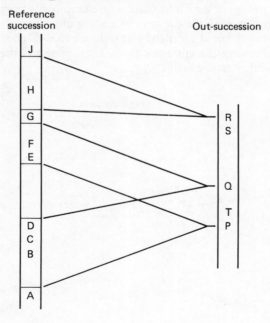

out-of-date methods mentioned above, and that the new proposals (see Section 14.6 above) in both paleontology and stratigraphy provide a simple approach by which improved resolution may be pursued as far as the data and the interest permit in any individual case. There remain, of course, many instances in which pursuit of refinement is uneconomic and not required, but it is highly desirable that there should exist a procedure in which all new data from whatever source should be brought to bear on problems that so merit such attention. Hitherto, no such choice was in general available.

Fig. 14.3B. Example of multiple bracket correlation in which the reference section is calibrated in metre-sized units from only some of which character-bearing samples have been obtained. Correlation can then be expressed positively if desired, rather than negatively as in Fig. 14.3A, e.g. P correlates with units 12–20, T with (12) to (23). Note that the characters of A–J and P–R can come from any fossil group, e.g. ammonoids, foraminifera, dinocysts, or from non-fossil characters, in any combination.

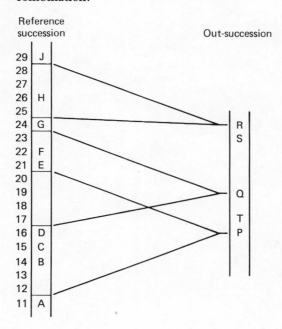

Part D. Further considerations

15

Human and other problems

A number of minor problems arise from consideration of the main issues under discussion above. They are not themselves part of the central argument, nor are they susceptible to logical ordering.

15.1   *Authority*. It has been suggested that it is presumptuous to publish without wide consultation an apparently iconoclastic scheme for handling primary information. Experience, however, over more than 25 years with the Botanical Code (ICBN) has been that only amendments to the Code with a narrow biological or a legalistic purpose could be contemplated; arrangements with a geologic or a wider general purpose were simply not understood enough to achieve serious discussion. Informal discussion on the zoological side (ICZN) about animal fossils indicated even greater rigidity. Since there has also been strong editorial bias towards complete acceptance of both Codes as central and mandatory in the field of fossils, it appears to be necessary to break with tradition. The only international organisation in this field, the International Palaeontological Association (IPA, originally IPU), has no mechanism and insufficient cohesion to accommodate such discussion; even as a member of its executive for a spell of four years I was quite unable to initiate within its scope any consideration of such matters at a Congress or otherwise. There is no other forum for paleontology as a whole, from which any authority might be sought; consequently normal scientific publication channels have to be used. The challenge therefore of PDHC is to advocate the restriction of both ICBN and ICZN to Holocene/Recent material, and also to recommend the re-writing of the *International Stratigraphic Guide* (Hedberg 1976) to about a quarter of its previous length and the removal of all reference therein to 'classification'.

15.2   *Use of terms.* Several of the concepts considered in the chapters above are new, although they are usually relatable individually to some existing concept. Choosing a name for such a new concept raises a dilemma; the use of several entirely new terms together produces an unfamiliar language which is counter-productive in achieving understanding, but the qualification or modification of existing terms (however seldom they have been used in the past) leads to an uncertainty of precision unless the new definition of the term is continuously available. Some examples will illustrate the difficulties.

First, an example may be taken from the discussion of the base-taxon known from the beginnings of paleontology as a species, despite the fact that this term was merely borrowed from neontology without any accompanying statement that its use for fossils was imprecise in not meeting the accepted definition for a 'species' of living organisms. The term 'biorecord' was coined two decades ago (Hughes and Moody-Stuart 1969) for an unalterable reference record, designed to take the place of a 'species of fossils'; experience has shown that the form of this selected name did not help enough in making the concept understandable. Almost the same concept is here named a 'paleotaxon' in Chapter 6 above, with the hope that this new but explicit term will prove more successful in making the purpose clear.

Secondly, at about the same time (Hughes 1969), it was necessary to coin a term for a new kind of 'genus' which was to be explicitly restricted by scope in morphology and in stratigraphic range. The word 'genusbox' indicated the parallel to a genus but maintained a distinction; it was more easily understood than the biorecord even by those who rejected the concept. This concept has now evolved (Chapter 7 above) to the state in which the equivalent of a genus bears no reference, implied or explicit, either to morphology or to biologic classification. The term 'timeslot' has been given to mark the purely stratigraphic basis for such a unit; the concept is entirely new and so therefore is the term.

Thirdly, in the stratigraphic field, the term 'event' was needed (Chapter 14 above) to mark an interpretative concept based on a single sample (of any size) from a single geologic succession. Although the term 'event' was scarcely used 20 years ago it has rapidly come to bear the additional meaning of a concept raised from a number of samples in different successions believed to be synchronous but in fact requiring correlation to prove this. In this case, although the term has become downgraded in this way to include something imprecise, a solution appears to lie best in the use of a qualifying prefix, e.g. palyn-event. Such a prefixed term is new, but unless its single sample definition is adhered to, it can become downgraded in the same way as the unprefixed term. No other short English, French, German or Latin word appears to be available.

The difficulties so encountered in these cases are surprising, although it is possible that they mostly arise because of the unwritten requirement for brevity in such names.

15.3 *Recalcitrant geologists.* Some critics fear that even a small number of geologists not wishing to subscribe to such a new scheme could render it unworkable. In practice, assuming that such colleagues or organisations adhere to past custom rather than embark on other distinct new schemes, there would be no difficulty. The procedures proposed here are all parallel and compatible with traditional procedures, and the distinction will be clear from the terms used; the better procedures for any selected purpose will prevail in the long run. The less good procedures will also automatically be labelled as such by their form and by their lack of detail.

15.4 *Use of past data.* An obvious source of concern with a new scheme is that earlier work might become wasted through inaccessibility. Because of the compatibility of new procedures with the old there is no basis for such concern; older records can easily be re-assembled from the literature into the new form, and graded comparison with a traditional species can be made in the same manner as with a new paleotaxon. The only problem likely to arise is that some older records will prove on examination to lack information in certain categories so that the new record forms may contain blank fields. Although this may appear to be damaging, the only new effect will be to pinpoint deficiencies that already exist; it will not create any gaps.

Systematic entering of old records in the new form will perhaps seldom be profitable. It will always be preferable only to translate records on demand, but a small allowance of time will at first be necessary for such activity.

15.5   ***The nature of the paleotaxon.*** Unlike a species of fossils which is deemed to have a duration or range in geologic time, a paleotaxon is based on only one sample and thus has no 'range'. A paleotaxon, however, plus all the CfA records compared with it, can if desired be said to have a range and can form therefore a parallel to the traditional species.

15.6   ***The acme problem with fossils.*** Especially with microfossils a single abundant occurrence of a taxon is inevitably an attraction. Unfortunately, it is very likely that the causes of individual acme occurrences are normally local and to that extent accidental. It is therefore only at the crudest level that acme occurrences can be used for correlation of two successions.

In PDHC work, however, an acme provides an excellent opportunity to produce strongly supported details of variation of a population from a single sample. Around this, good comparison records from other occurrences can be built until as fine a correlation bracket as desired has been achieved.

15.7   ***The scope of a 'palyn-event'.*** If the event under study is the first appearance of palynomorph P, and the traditional species and attributed specimens are used, the usual problem is whether or not to score single specimens found in samples below the first well-supported occurrence (Fig. 15.1). Rules can be drawn which exclude either one single specimen or even only two specimens as not significant and therefore unscored; but if a larger fossil such as a megaspore or a distinctive seed megafossil is used, failure to score becomes debatable and difficult to sustain. Using PDHC, all that is necessary with any type of fossil is to agree beforehand what size of assemblage of specimens will be acceptable as significant in a CfA comparison record (Fig. 15.1); this is selected here to be 25% of the number of specimens used to establish the relevant paleotaxon. This procedure does not change any array of facts, but does provide criteria of the same quality for use in two or more successions.

15.8   *Methods of dissemination of data.* Although the GOR form (Chapter 4 above), the PTR form (Chapter 6) and other illustrations have been designed with electronic storage and transmission in mind, they are equally suitable for regular publication in journals and would lend themselves to camera-ready preparation (see Penny 1988). In particular, because of the originator/date/time designation, there is no need for any 'umbrella' control organisation of such data sources.

Fig. 15.1. Diagram to illustrate the problem of significance of a single specimen occurrence of a palynomorph or other microfossil. Application of a convention that 25% of numbers used for a paleotaxon will suffice for recording CfA, leaves 1333 as a CfA record but 1394 not (cf., ungraded). Such a convention could be varied to accommodate different purposes.

	Borehole depth samples	Numbers of specimens	Record
	1298	10	Barremian-croton (Paleotaxon)
	1316	0	—
Accepted	1333	3	CfA Barremian-croton
	1345	0	—
	1353	0	—
	1394	1	cf.-croton (ungraded)

Entirely separate consideration of present-day geologic and paleontologic journals suggests that the trend already is to eliminate or
to reduce where possible the formal publication of purely record or
systematic material. This reduction is advocated on grounds of (a)
expense to both institutional and private subscribers, and (b) un-
readability of such material to all but the few other research workers
who are obliged to read it (how often are scientific papers fully read
and by whom?).

Thus the proposals outlined above are deliberately aimed at prob-
able future methods of transmission, without any loss of contact with
the present.

15.9   *Storage and retrieval of data.* For simplicity it has been
          assumed above that records would all be stored by the orig-
inator's institution or by an institution designated by the originator.
This may be thought to overlook either the expense of the operation
or the impermanence of some institutions, but there is nothing to
preclude the take-over of collections and data by large national
institutions deemed to be permanent. It has been suggested that
monitoring and curation of the data-base will cost money; but I
would suggest that no monitoring beyond appointment of adequately competent staff will be necessary and this would apply to all
other activities. Once a record is made it is immutable, thus requiring no attention other than retrieval. The only curation necessary
would relate to the electronic equipment, and not to the nature of the
data. Ideally, record data would be equally available to all scientists,
but realistically this would be subject as now to the withering away of
governmental and commercial restrictions; perhaps eventually all
such restrictions will be held to strict and short time-limits by law.

Although stored data on observational records may be theoretically available from any institution on demand, the cost of retrieval
and transmission is significant. In any one nation it may well be
possible to arrange to ignore these costs on a basis of exchanging
benefits; across frontiers the process is cumbersome and expensive.
Is it too idealistic to suggest that improvements may be expected
because of equivalent pressure from many other subjects? It
also seems possible to imagine ultimately a reduction of 'security'

difficulties in the same way, but as for several decades already genuinely sensitive information will simply not be available to science. Such difficulties cannot be said to form any argument for failing to make scientific progress with this proposed scheme; the scheme would suffer from current restrictions in the same way as does traditional science, but would be ready to benefit from each and every easement of international incongruities.

16

Main plan re-stated

16.1   **Purpose.** The principal purpose throughout this volume has been to promote full appreciation and discussion of the many problems raised by the pursuit of traditional methods of study of paleontology and of stratigraphy. In particular I have tried to assess the severely qualified success so far of these methods in achieving greater resolution from application of more and more of the additional information constantly becoming available from advances in instrumentation and from better generally integrated understanding.

Although workable solutions are offered to all the problems so far recognised, their immediate acceptance is a less urgent matter than agreement on the analysis of the difficulties.

16.2   **Esssentials.** The main difficulties with present methods can be expressed in terms of four needs:
  (a) Equal treatment of all observational records.
  (b) Accurate expression of comparison of new material with existing records, allowing for continuing advances of knowledge.
  (c) Production of correlation statements designed for continuous refinement with advancing knowledge.
  (d) Arrangements for truly phylogenetic universal classifications flexible enough to absorb new information.

These points will be summarised separately below in the next four numbered sections. All other remaining points, including the tightening up of definitions of taxa and the provision of adequate but not excessive nomenclature, are considered as logical adjustments to the four essentials.

16.3   ***Equal treatment of all records.*** The reason for the long-standing toleration of inadequate paleontologic records is almost certainly historical. In the early days of exploration geology, it was manifestly quicker and more intelligible to label a new occurrence as attributable to an existing taxon if that was at all possible. Such attributions were then on the frontiers of knowledge. As time passed the results became either (a) a massive loss of data if the attribution was imperfect or (b) a cumbersome (but presented as scholarly) search through synonymies which had more to do with

Fig. 16.1. Diagram to compare custom and specific purpose in treatment of a microfossil assemblage.

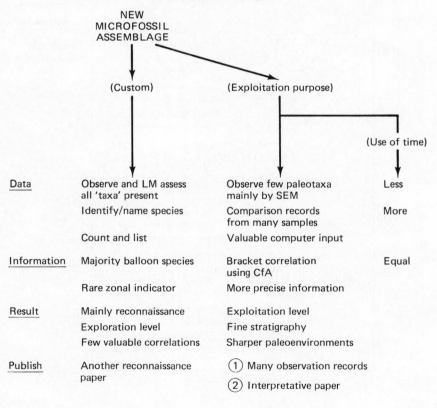

nomenclatural niceties than records. In the study of some megafossil groups good human brains have been devoted to managing the resulting chaos, which exercise was often just possible in a small group of fossils. The application of similar methods without thought to microfossil groups with almost unlimited data (Fig. 16.1) has left a

Fig. 16.2. Diagram to illustrate a simple case of three states of a microfossil species circumscription, all under the same name. 1965: Attributions A–D, frequently in part, are a factor in further attribution. 1970: Improved definition of species from more and better specimens; holotype persisting. 1975: New attributions E–G, but are A–D still included? 1980: New characters but usually not involving nomenclature holotype. 1985: New attributions J–L, but are A–G included in range or not? This account assumes rapid and efficient transmission and use of publications round the world.

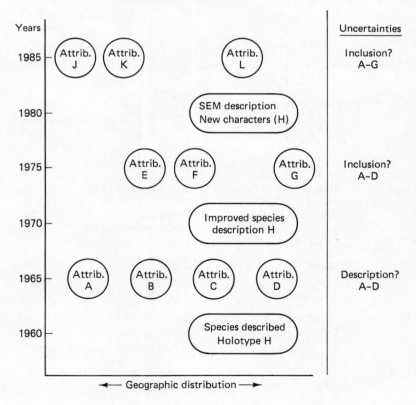

largely barren literature of often unusable fragments of information to which considerable additions are still being made. This is beyond the scope of single human brain for handling or for efficient rational use.

Clearly, basic data of observation records will have to go out of the published literature very soon, into data-storage and retrieval systems with at least professional access worldwide. Now that large data-sets can be searched systematically and the results manipulated, there is no longer any need to exercise taxonomic or nomenclatural parsimony for the supposed benefit of the human brain at work. Records can now be expressed all in the same way in adequate detail; as long as the expressed descriptions are formalised and also contain critical keywords for retrieval, the records can remain in the database in their original form. Any attempts at taxonomy, nomenclature or classification will be independent of any record retrieval system; such attempts can then be used or completely ignored without difficulty.

16.4   *Comparison, allowing for continuing advances of knowledge.* A species of fossils as originally described is a simple and acceptable concept; comparison with it is straightforward and uncomplicated. Attributions to a species immediately increase the scope of the target species by consolidation, so that each new attribution is compared with the consolidated species as then known to the attributing author. Consequently each comparison is made with a slightly different moving target (Fig. 16.2). The pace of movement of the target may then be quickened by emendation as a result of new discovery, of description or of definition of the species (which is normally continuing under the same name). In theory all these changes could be logged and each 'attributor' could accurately report on the scope of the moving target, but this does not happen all over a far-flung real world. In practice the existence of such a problem is often denied because the effects being gradual and continuous make little difference in a short period of observer-time.

The solutions offered are (a) to insist on a graded comparison of character variation-state in the new record with the character variation-state in the original species record (this is slightly more difficult to carry out and must be done by the author while he or she is actually working on the material), and (b) erection of a new record with a new designation (and also new name) in the case of any emendation whatever.

This solution certainly produces a greater volume of records but each one is a valuable statement even in isolation from all other records. It can therefore be retrieved from data-storage alone, and used without reference to any other taxon or record.

16.5   *Correlations designed to be improved by continuous refinement.* Stratigraphic time-correlations have by tradition been made, after considering all available evidence, as spot decisions of a best-fit guess nature. These are based of course on all past information available, but make no allowance for useful incorporation of new information received tomorrow; this despite the fact that while scientific objectives are pursued new information and interpretations will always be added in the future.

The scheme offered here is to make all correlation statements in pairs, each of a stated probability, forming a bracket (Fig. 16.3) which remains continuously capable of refinement on receipt of new information. Thus all correlations will remain open to some extent, but on an ever-decreasing scale; they will also accurately reflect the state of knowledge at the time of each statement.

Further, in a pair of statements making a bracket, the two statements can be independently based on quite separate criteria. In other words, the way is open to use all data, from any kind of fossil or other phenomenon, together without any declaration of zones or other units.

16.6   *Towards a universal classification.* Classification of very imperfectly known fossil organisms with their relatively fully understood Holocene descendants has always led to classification taxa entirely biassed in favour of the latter (Fig. 16.4A). This has led to simple assumptions about the presence of organs in past organisms, notably the 'soft-parts' of invertebrate animals, far beyond the evidence of the fossils themselves (Crowson 1970, p. 67). In an era of special interest in paleoenvironments as such there has been a strong tendency to use the 'evidence of past organisms' on the slenderest pretext.

The suggested arrangement is to have many successive classifications of all the fossils known in each period of geologic time without admitting any other information (Fig. 16.4B) and in

Fig. 16.3. Example of development of correlations of an out-succession sample X with a reference borehole succession. A first correlation bracket is established on the evidence shown (P, on the right) – first at B̊H. Another paleotaxon (Q on the left) can then be used independently to refine the bracket to E̊H only. Further improvement would probably require denser sampling of the reference succession. (Graded comparison record CfA requires 25% of the specimens employed in the variation of the paleotaxon.)

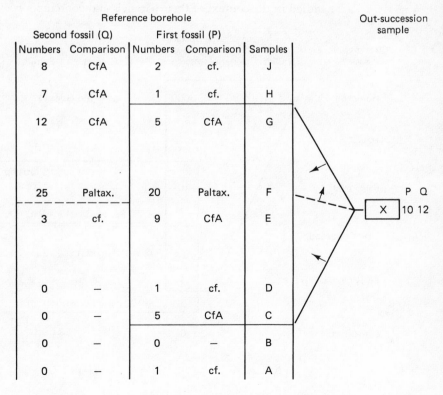

Reference borehole					Out-succession sample
Second fossil (Q)		First fossil (P)			
Numbers	Comparison	Numbers	Comparison	Samples	
8	CfA	2	cf.	J	
7	CfA	1	cf.	H	
12	CfA	5	CfA	G	
25	Paltax.	20	Paltax.	F	P Q
3	cf.	9	CfA	E	X 10 12
0	–	1	cf.	D	
0	–	5	CfA	C	
0	–	0	–	B	
0	–	1	cf.	A	

particular by restricting knowledge of living organisms to the Holocene classification alone. These 'Period classifications', or parts of
them, could then be strung together to approach a historical description of each type of fossil through enlarging (or occasionally contract-
ing) groups. At first sight the elimination of information not actually
derived directly from the fossils would apparently reduce the supply
of usable data and so perhaps jeopardise interpretations in such
interpretations of paleoenvironments, but the pay-off would come
later when the much more restricted deductions possible would be
seen to be very much more soundly based and reliable.

Fig. 16.4. Diagram to illustrate and compare the effects of (A)
'Traditional classification' in which a small number of fossils
(very incomplete organisms) tend to be lost and poorly
compared among a very large number of living species, and (B)
'Period classification' in which the records of fossils of parts of
organisms (e.g. pollen and leaves) may be considered and
handled in the context of their parallels and contemporaries.

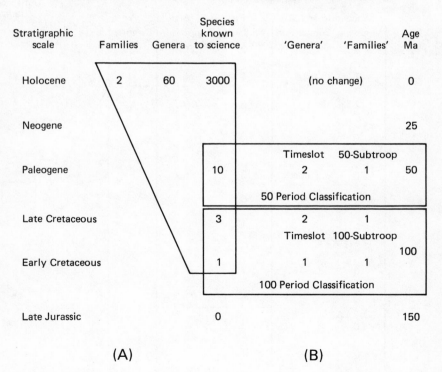

Stratigraphic scale	Families	Genera	Species known to science	'Genera'	'Families'	Age Ma
Holocene	2	60	3000	(no change)		0
Neogene						25
Paleogene			10	Timeslot 50-Subtroop 2	1	50
				50 Period Classification		
Late Cretaceous			3	2	1	
				Timeslot 100-Subtroop		100
Early Cretaceous			1	1	1	
				100 Period Classification		
Late Jurassic			0			150

(A)   (B)

16.7 ***Artificial intelligence.*** Proposals from Chapter 10 onwards
are intended to be suitable for computer handling, through
the non-numerical 'artificial intelligence' languages such as 'Prolog'
(Riedel and Tway 1989). These languages can be regarded as
developing 'expert systems' to amplify greatly the work of the human
brain rather than in any sense providing the necessary insight
independently.

16.8 ***Other possible solutions.*** It is quite possible that other plans
will prove to be better than the suggestions made above.
Most desirable first, however, is proper scientific debate on the
validity of the analysis and it is hoped that the labelling of the
essentials in Section 16.2 above may make such discussion easier to
achieve.

APPENDIX 1

Example entry of individual comparison record of a Cretaceous pollen to illustrate the use of a PDHC General Observation record Form (GOR); all of the information placed to the right of the double vertical line can be transmitted as written under the field letters in order (A–T).

The design of the GOR form is intended to remind the user and to make available all the necessary information in a logical order. Although it is reproduced at A5 size in this book, it is intended for normal use at full A4 size. Any serious use of such a form is expected to lead to its further development for greater convenience; for example, the left-hand side printing could occupy much less space, and vertical space proportions of some of the fields could be varied.

Detailed comment on this example:
  H Numerical age is rarely available; paleomagnetic data more frequently.

J/K Compatibility arrangement allows use of traditional Linnean taxa as necessary.

  M Although it is intended that illustration figures will be entered on the same form (M1–6), it is still usually necessary to transmit them separately and they have therefore been omitted from the reproduction of this example.

Main identifier	Originator	A	(Name) McDougall, A.B. (Address) Earth Sciences, Cambridge University	
	Date/time	B	(year-F) 1987 (month) October (day-F) 6 (hour)	
Rock sample	Locality	C	(Grid) Brit. Nat. (Co-ords) SZ440807 ± +2	Fig. M1
	Rock formation	D	Vectis Formation, Cowleaze Chine Member	
	Sample position	E	BRN 016, mid - Bed 35 (White 1921) prep. X354, stub AM320, film 02	
	Sample lithology	F	Grey sandy clay	
	Succession age	G	(Era) Mesozoic (Period) Cretaceous (Age/stage) Barremian (Chron)	Fig. M2
	Numerical age	H	Ma: — ± — Ma. N (PMag)	
Fossils	Paleobiogroup	J	Early cret poll	
	Graded comparison	K	CfA (Timeslot/genus) Superret (Paleotaxon/species) croton Hughes et al. 1979	Fig. M3
	Number of specimens	L	30	
	Description	M	Biorecord description pp. 520 + 524. Diameter 25 (29) 39 μm. Supratectal rectangular plates, one to four, between triangular plates. Granules (max. diam. 0·2 μm) on distal surface of supratectal plates.	Fig. M4
	Variation	N	Size difference Supratectal granules	Fig. M5
	Preservation	P	Palyn. good	
	Facies	Q	Palyn. Retichot-baccat, Retisulc-dentat, Superret-triang, 'Ephedripites'	
Reference	Specimen repository	R	Hughes Colln., Sedgwick Museum Cambridge, UK.	Fig. M6
	Earlier records	S	Drewry, G.E., Earth Sci. Cambridge 1979 Sept 14; stub GD317, film B264	
	Record ends	T	(ends)	

APPENDIX 2

An example of published use of the PDHC Paleotaxon Record Form (PTR) is taken from Penny (1988, p. 381). This was based on an earlier (1986) version of the form, but the main purpose and the sequence of fields are the same. Although PTR forms may be submitted for formal publication more frequently than examples of the General Observation Record (GOR), it is expected that future use of both will be mainly for transmission between geologists as outlined in the text.

Detailed comment on this example:

A   Mention of Earth Sciences in the University is necessary.

B   The date/time should read '1987 July 8 fifteen 36', in logical order without commas.

M   The illustrations were separately reproduced on Plate 29 (p. 383) and are not copied here.

C   Height of borehole (RTE) is not given.

TABLE 2. Biorecord: RETIMONO-SPINEROW.

REFERENCE TAXON DESCRIPTION

Group of organisms	J	Monocolpate pollen.
Sequence age	G	Mesozoic/Cretaceous/late Aptian.
Originator	A	Penny, J. H. J. Cambridge University, UK.
Origination date	B	1987, 8 July, fifteen 36.
Taxon name	K	Biorecord: RETIMONO-SPINEROW.
Description	M	(All observations made with SEM.)

Monocolpate pollen, rounded outline, maximum diameter 14·5(17·5)20·0 μm. Exine semitectate, reticulate, lumina rounded to irregularly polygonal with even size distribution, maximum internal diameter 0·9(2·0)3·1 μm. Microlumina absent. Muri rounded in cross-section or slightly wider than tall, height 0·4(0·5)0·6 μm, width 0·5(0·7)0·9 μm; upper surface of muri sculped with distinct spines up to 0·4 μm tall and often arranged in pairs on opposite sides of the muri, bases of spines may be united to form transverse ridges; sides and lower surfaces of muri unsculped, columellae absent. Aperture long, up to half the circumference of the grain, margins continuous, unspecialized with no tendency for adjacent lumina to be smaller than on the main body of the grain; there is a corresponding slit-like aperture in the nexine. Nexine smooth, sometimes closely applied to the sexine but usually separated by a distinct gap. Accurate nexine measurement obstructed by sexine, range 12·9–20·0 μm. Nexine may be rotated inside the sexine.

Variation record	N	Recorded under M.
Number of specimens	L	9.
Locality	C	Mersa Matruh borehole, N.-W. Desert, Egypt. Grid ref. 31° 19′ 43.00″ N. 27° 16′ 07.00″ E.
Rock formation	D	Kharita.
Sample position	E	MMX-1 7890, at depth 7890 ft.
Sample lithology	F	Fine-grained yellow sandstone.
Preservation	P	Good.
Repository	R	Dept. Earth Sciences, Cambridge University, UK. Preparations JP 066, 180. Stubs JPS 228, 229, 230.
Earlier records	S	None.
Conclusion	T	Ends.

GLOSSARY

The second column of abbreviated subjects names indicates the appropriate subject field, either to assist the reader or to distinguish from any other possible usage.

The small number of terms underlined are those which are newly introduced in this work or have been introduced by the author in the last 20 years as part of these proposals.

adaptation	(biol.)	refers to the suitability of the morphology of an organism to the presumed function
ADP		automated data processing
Age (stage)	(geol.)	term for a formal division of a 'period' in the stratigraphic time-scale. 'Age' is correct; 'stage' is commonly used
angiosperm	(biol.)	name for the largest group of living land seed-plants; many Tertiary and some Mesozoic fossils may be included
anthropocentric		centring in man, or seen from a human point of view
assemblage	(paleont.)	a set of fossils selected from a sample of rock, or a group of samples
assemblage-zone	(strat.)	a zone characterised by a number of taxa (of different ranges) in their occurrence in the same assemblage
attribution	(paleont.)	placing of fossil specimens from other samples into a species, thereby extending the meaning of the species beyond that of the original occurrence

balloon taxon	(paleont.)	a taxon of fossils which has grown greatly in scope through attribution of many other occurrences of specimens
base-taxon	(biol.)	the first or originally erected taxon for a kind of organism or of fossil, traditionally a species
binominal	(biol.)	the two-element style of name (e.g. genus + species) on which the existing Nomenclature Codes are based; or another two-element construction
biologic evolution	(biol.)	sector of earth evolution describing the successive occurrences of past organisms
biorecord	(paleont.)	term for an unmodifiable base-taxon of fossils (Hughes and Moody-Stuart 1969); now superseded, see paleotaxon
biostratigraphy	(strat.)	superfluous classificatory term intended to include all stratigraphic work employing fossils
biozone	(strat.)	zone characterised entirely by fossil remains of organisms
boundary-stratotype	(strat.)	short rock succession, selected for its correlation potential, to contain a marker-point (or GSSP)
bracket correlation	(strat.)	pair of succession-correlation statements, capable of continuous refinement when using new information
calibration	(strat.)	graduation of a stratigraphic scale boundary (or other) section with observation records for reference
cf. (confer)	(biol.)	symbol of comparison with a base-taxon, indicating less positive statement than attribution
CfA	(paleont.)	comparison record A, positively indicating close similarity to a reference base-taxon
CfB	(paleont.)	comparison record B, negatively indicating one qualitative difference from specimens of a reference base-taxon
CfC	(paleont.)	comparison record C, indicating open record of some similarity with a stated base-taxon but likely to

		represent comparison with another as yet undescribed new taxon eventually
character	(biol.)	a quality of an organism, or of a fossil
Chron	(strat.)	small division of an 'Age' in the stratigraphic time-scale
chronostratigraphy	(strat.)	superfluous classificatory term for so-called time–rock units (systems, series, stages, chronozones)
chronozone	(strat.)	superfluous term for the rocks produced in the duration of a chron
clade	(biol.)	a set of taxa interpreted as being derived from a single common ancestor
cladistics	(biol.)	an analytical study of character-states of taxa aimed at clarifying relationships
class	(biol.)	level in bioclassificatory hierarchy between order (below) and division/phylum (above)
classification		arrangement of material for ease of discussion or retrieval
cluster		group of similar things, which grows naturally in scope as more are recognised
comparison record	(paleont.)	record of fossil specimens, related by the original author of the record to a base-taxon and labelled CfA, CfB or CfC
concurrent-range-zone	(strat.)	zone marked by the overlap of ranges of two or more fossils
consecutive-range-zone	(strat.)	zone with boundaries marked by mutually exclusive types of fossil in the same lineage
correlation	(strat.)	relation between two things, in this case in time of formation between selected parts of two rock successions
data-handling		arangement of conventions that are necessary to bring observational records to bear on geologic and paleobiologic interpretation
date/time		record of 'geologist' time-instant in terms of year, month, day, hour, minute, in that order
definition	(strat.)	stating the precise nature of a thing, e.g. of a base-taxon of fossils of organisms
descent		ancestry

diachrony	(strat.)	the observation or deduction that a recorded stratigraphic feature observed in more than one place may differ in age in its separate occurrences
dichotomy		division into two by repeated bifurcation (as of veins of a leaf)
dinocyst	(biol.)	the resting cyst of an organic-walled aquatic planktonic organism, living or fossil
division	(biol.)	a level in bioclassificatory hierarchy for plants only between class (below) and kingdom (above); equivalent of 'phylum' for animals
division	(strat.)	a sector of the stratigraphic time-scale, between two described marker-points in rock
earth evolution		interrelated changes with time in the earth as a whole, or as observed in parts of the earth
emendation	(biol.)	formalised change to a previously formal name or definition, notionally removing errors or adding characters
ephemera		things of short-lived or transitory value
Epoch	(strat.)	term for a formal division of the stratigraphic time-scale between 'Age' (shorter) and 'Period' (longer)
event	(geol.)	an interpretation raised by an observer from a rock sample of any stated size; usually qualified by a descriptive prefix
event-correlation	(strat.)	correlation effected by the use of separate events rather than through the use of zones
evolution	(general)	progressive change, with the new state consequent on the nature of the old
exploitation	(geol.)	the integrated development of knowledge to achieve more-precise interpretation
exploration	(geol.)	the investigation of geologic phenomena, not previously studied in the same detail
extant	(biol.)	of organisms existing in Holocene time or still living

FAD	(geol.)	First Appearance Datum. Age of rocks in which specified fossils are earliest known
family	(biol.)	level in bioclassificatory hierarchy between genus (below) and order (above)
floral/faunal list	(strat.)	a list of occurrences of fossils in an assemblage, by taxon; sometimes quantitative
genetics	(biol.)	the study of heredity and variation
genus	(biol.)	the first aggregation of species into the lowest level of bioclassificatory hierarchy, below 'family'
genusbox	(paleont.)	a deliberately restricted kind of genus (Hughes 1970); now superseded by 'timeslot'
geochronology	(strat.)	superfluous classificatory term for time determination
geochronometry	(geol.)	a technique for making radiometric time determinations
GOR	(new)	General Observation Record. Form of presentation of data for all except base-taxa (see PTR)
graded comparison	(paleont.)	a formally stated comparison of an observation record with a base-taxon in terms of CfA, CfB or CfC
GSSP	(strat.)	officially designated scale reference point (marker-point) on the Global Stratigraphic (Reference) Scale
hierarchy		a graded organisation
Holocene	(geol.)	latest epoch of the stratigraphic time-scale, commencing 10 000 years ago
holotype	(paleont.)	a single fossil specimen, selected as a permanent nomenclatural reference, from material used in the original description of the base-taxon concerned
100-subtroop	(new)	a unit in a 'Period classification' containing Cretaceous fossils only, based on a 100 Ma time-plane (Cretaceous-Albian)
ICBN		International Code of Botanical Nomenclature (Greuter *et al.* 1988)
ICZN		International Code of Zoological Nomenclature

identification	(biol.)	statement of similarity of new specimens to an established taxon, testable in living organisms but not in fossils
IPA		International Palaeontological Association
isochronous	(geol.)	of two or more geologic phenomena from separate samples assumed to be of the same time of origin; very rarely capable of proof
LAD		Last Appearance Datum
lineage	(paleont.)	ancestry, usually expressed in terms of base-taxa believed to be successors in time
lineage zone	(geol.)	zone based on a taxon of fossils, believed to be directly descended from those characterising the previous zone (= consecutive-range-zone)
Linnaean	(biol.)	of taxa with formal binominal nomenclature, using ICBN, ICZN or other such Codes
lithologic	(geol.)	of all the properties of rocks
lithostratigraphy	(strat.)	classificatory term which is best dropped in favour of 'rock description' because of a dispute about its scope of meaning
LM		Light Microscope
lumping	(paleont.)	of a style of taxonomy of fossils, economical with the numbers of taxa erected; see also 'splitting'
macro-evolution	(biol.)	evolution expressed in terms of higher taxa (families and above) related to each other in succession or in contemporaneity: validity of concept uncertain
marker-point	(strat.)	selected and designated point in a rock succession, employed to mark the beginning of a stratigraphic time-scale division
megafossils	(paleont.)	fossils which are large enough to study without a microscope
megaspores	(paleont.)	in certain spore (not seed) plants, large spores (usually 200 μm to 2 mm diameter), notionally carrying the female gametophyte

mesofossils	(paleont.)	term applied to fossils of intermediate size, in particular to pieces of plant cuticle and wood, and megaspores
microfossils	(paleont.)	fossils for the study of which some form of microscope is essential
morphologic	(paleont.)	of the form of plant and animal fossils
natural selection	(biol.)	selection for eventual breeding, by the effect of outside agencies, of certain juvenile organisms from an original large production
neoteny	(biol.)	supposed prolongation of larval or juvenile development stages into adult life
nomina conservanda	(biol.)	names of taxa retained by agreement in cases in which strict application of ICBN/ICZN Rules would invalidate them
Oppel-zones	(strat.)	assemblage-zones characterised in terms of the ranges of numerous fossils
order	(biol.)	level in bioclassificatory hierarchy between family (below) and class (above)
organ-taxon	(paleont.)	taxon of fossils based on an organ, such as a leaf or seed, without knowledge of the complete organism
originator	(paleont.)	the person describing an observational record for the first time
orthographic	(biol.)	of correct or conventional spelling of names
out-succession	(geol.)	rock succession, newly under consideration, for comparison with a described reference succession
palaeontology	(geol.)	the study of fossils; preferred spelling in Europe; see also paleontology
paleobiogroup	(new)	major morphologic and non-hierarchic grouping of types of fossil
paleobotany	(paleont.)	the study of fossils of plants
paleoentomology	(paleont.)	the study of fossils of insects
paleoenvironment	(geol.)	reconstruction of a past geologic environment
paleomagnetic	(geol.)	of (fossil) remanent magnetism of rocks
paleontology	(geol.)	the study of fossils; shortened spelling used here

paleopalynology	(paleont.)	the study of fossils of spores, pollen and organic microplankton
paleotaxon	(new)	a new form of immutable base-taxon of fossils; developed from a biorecord
palingenesis	(biol.)	exact reproduction of an ancestral character in later evolutionary succession
palynomorph	(biol.)	a very small independent organic-walled body, e.g. pollen
parataxon	(paleont.)	a taxon of fossils based on something less than a whole organism
PDHC	(new)	Paleontologic Data-Handling Code
Period	(strat.)	a large division of the stratigraphic time-scale, e.g. the Cretaceous Period
Period classification	(new)	a new classification of fossils confined to records from a named geologic Period, indicated by a numerical age prefix, e.g. Cretaceous Albian Period classification = 100-classification
phylogeny	(biol.)	history of evolution or descent of an organism type
phylum	(biol.)	level in bioclassificatory hierarchy of animals between class (below) and kingdom (above); equivalent to 'division' for plants
population	(biol.)	all members of a species present at any one time and/or in any one place
pre-adaptation	(biol.)	a morphologic character already developed in an organism prior to its functional employment
Psychozoa	(biol.)	the most developed animals, deemed to have some influence on their evolutionary fate
PTR	(new)	Paleotaxon Record. Form of presentation of data for base-taxon only
publication	(biol.)	dissemination of published information is described as 'effective' or not for the Rules of ICBN/ICZN
punctuated equilibrium	(paleont.)	a supposed evolutionary pattern in which evolution proceeds rapidly in short bursts
radiation	(biol.)	evolutionary pattern of very rapid appearance of many diverse descendants of one lineage
radiometric	(geol.)	of age determinations in years, derived from study of the products of radioactive decay

range	(paleont.)	duration of occurrence of a specified kind of fossil
record	(geol.)	statement of occurrence of any fossil or other geologic phenomenon
repository	(geol.)	institution holding specified geologic material
rock description	(strat.)	the fundamental first discipline of stratigraphy, embracing all characters observed in rock
rock formation	(geol.)	unit composed of rock strata, erected to aid description of extensive successions
sample	(geol.)	collected or observed rock material of any stated scope or size
SEM		Scanning Electron Microscope
species	(biol.)	traditional base-taxon, composed of living individuals capable of interbreeding; 'species' of fossils cannot depend on this criterion
splitting	(paleont.)	a style of taxonomy of fossils in which small differences are taken to distinguish new taxa; see also 'lumping'
stage	(strat.)	superfluous term for all the rocks laid down in the duration of an 'Age' of the stratigraphic time-scale. Commonly used with the same meaning as 'Age'
stratigraphy	(geol.)	the descriptive study of rock strata (layers); also the interpretation of geologic history through the descriptive study
stratology	(geol.)	little-used name for the interpretation of geologic history; its use would separate this meaning from the descriptive study which is correctly named 'stratigraphy'
subtroop	(new)	a term for a unit in a 'Period classification', incorporating about 100 paleotaxa or species (see troop)
succession	(geol.)	describes a series of rock layers (strata) subject to the Law of Superposition which states that the younger rock lies above the older
succession age	(strat.)	relative age of a sample in relation to other rock samples
taphonomy	(paleont.)	reconstruction of the mode of burial and diagenesis of fossils from study of preservation of fossil and sediment

taxon	(biol.)	any group of organisms with a stated similarity; plural 'taxa'
taxonomy	(biol.)	the art of distinguishing separate taxa
time-correlation	(strat.)	widely used to mean succession-correlation because successions themselves indicate passage of time
time-level	(strat.)	indicates a selected time (marked in rock) in one succession and its supposed equivalents elsewhere
time-slice	(strat.)	indicates time between two time-levels
timeslot	(new)	new substitute for the genus taxon consisting only of the name of a stratigraphic time-scale division
topotype	(paleont.)	a specimen of a fossil taken from the sample from which a base-taxon has previously been characterised
trinominal	(new)	a formal three-element name-set for a fossil in PDHC, i.e. paleobiogroup-timeslot-paleotaxon
troop	(new)	a term for a large unit in a 'Period classification', incorporating notionally about 1000 paleotaxa or species (see subtroop)
type locality	(paleont.)	sample locality from which was collected the holotypic specimen of a species, or the whole material of a paleotaxon
unconformity	(geol.)	a break of kind or of inclination between adjacent strata, which indicates a hiatus of non-deposition
uniformitarian	(geol.)	of the view that past geologic processes have remained the same; a misleading simplification
uninominal		the name of a base-taxon consists of one word until the name of a genus or genus equivalent is added to aid filing as a binominal
variation	(biol.)	the recorded extent of differences in the characters of individuals of a species of organism, reflecting genetic dissimilarities in individuals
zone	(geol.)	generally understood as a unit of (usually stratified) rock characterised by a geologic phenomenon, of which the name is used as a prefix to -zone

REFERENCES

BERGGREN, W. A. and VAN COUVERING, J. A. 1978. Biochronology. In COHEE, G. V. *et al.* (Eds.) pp. 39–55.

COHEE, G. V., GLAESSNER, M. F. and HEDBERG, H. D. (Eds.) 1978. Contributions to the geologic time scale. *American Association of Petroleum Geologists, Studies in Geology* **6**, 388 pp.

COWIE, J. W., ZIEGLER, W., BOUCOT, A. J., BASSETT, M. G. and RE-MANE, J. 1986. Guidelines and Statutes of the International Commission on Stratigraphy (ICS). *Courier Forschungs–Institut Senckenberg*, Frankfurt a.M. **83**, 1–14.

CROWSON, R. A. 1970. *Classification and Biology*. Atherton Press, New York, 350 pp.

CUBITT, J. M. and REYMENT, R. A. (Eds.) 1982. *Quantitative Stratigraphic Correlation*. Wiley, New York, 301 pp.

DOYLE, P. S. and RIEDEL, W. R. 1979. Cretaceous to Neogene ichthyoliths in a giant piston core from the Central North Pacific. *Micropaleontology* **25**, 337–364.

FAHRAEUS, L. E. 1986. Spectres of biostratigraphic resolution and precision: Rock accumulation rates, processes of speciation and paleoecological restraints. *Newsletters in Stratigraphy* **15**, 150–162.

FORDHAM, B. G. 1986. Miocene–Pleistocene planktic foraminifers from D.S.D.P. sites 208 and 77, and phylogeny and classification of Cenozoic species. *Evolutionary Monographs* **6**, 200 pp, 25 pls.

GLUZBAR, E. A. 1987. Trustworthy biostratigraphic information is the only way to promote computerized information-handling systems in palaeopalynology. *Review Palaeobotany Palynology* **52**, 131–135.

GRADSTEIN, F. M., AGTERBERG, F. P., BROWER, J. C. and SCHWAR-ZACHER, W. S. 1985. *Quantitative Stratigraphy*. D. Reidel Publishing Co., Dordrecht and Unesco, Paris, 598 pp.

GREUTER, W., McNEILL, J. *et al.* (Eds.) 1988. International Code of Botanical Nomenclature. *Regnum Vegetabile* **118**, xiv + 328 pp.

HARLAND, W. B. 1977. International Stratigraphic Guide, 1976. Essay Review. *Geological Magazine* **114**(3), 229–235.

HARLAND, W. B., COX, A. V., LLEWELLYN, P. G., PICKTON, C. A. G.,

SMITH, A. G. and WALTERS, R. 1982. *A Geologic Time Scale.* Cambridge University Press, Cambridge, 131 pp.

HAY, W. W. 1972. Probabilistic stratigraphy *Ecologae Geologicae Helvetiae* **65**, 255–266.

HEDBERG, H. D. (Ed.) 1976. *International Stratigraphic Guide.* International Subcommission on Stratigraphic Classification of IUGS Commission on Stratigraphy, John Wiley & Sons, New York, 200 pp.

HOLLAND, C. H. 1986. Does the golden spike still glitter. *Journal of the Geological Society, London* **143**, 3–21.

HUGHES, N. F. 1969. Suggestion for better handling of the genus in Palaeopalynology. *Grana palynologica* **9**, 137–146.

HUGHES, N. F. 1970. The need for agreed standards of recording in Palaeopalynology and Palaeobotany. *Paläontologische Abhandlungen* **3B**, 357–364.

HUGHES, N. F. 1976. *Palaeobiology of angiosperm Origins.* Cambridge University Press, Cambridge, 242 pp.

HUGHES, N. F. 1978. Proposals for a new palaeobotanical appendix for the International Code of Botanical Nomenclature. *Taxon* **27**, 497–504.

HUGHES, N. F. 1986. The problems of data-handling for early angiosperm-like pollen. In SPICER, R. A. and THOMAS, B. A. (Eds.) Systematic and taxonomic approaches to palaeobotany. *Systematics Association Series* **31**, 233–251.

HUGHES, N. F., DREWRY, G. E. and LAING, J. F. 1979. Barremian earliest angiosperm pollen. *Palaeontology* **22**, 513–535, 8 pls.

HUGHES, N. F. and MOODY-STUART, J. C. 1967. Proposed method of recording pre-Quaternary palynological data. *Review Palaeobotany Palynology* **3**, 347–358.

HUGHES, N. F. and MOODY-STUART, J. C. 1969. A method of stratigraphic correlation using early Cretaceous miospores. *Palaeontology* **12**, 84–111, 10 pls.

INTERNATIONAL SUBCOMMISSION ON STRATIGRAPHIC CLASSIFICATION (A. SALVADOR, CHAIRMAN) 1987. Unconformity-bounded stratigraphic units. *Geological Society of America, Bulletin* **98**, 232–237 (Discussion in Volumes 99 and 100).

MATTHEWS, S. C. 1973. Notes on open nomenclature and on synonymy lists. *Palaeontology* **16**, 713–719.

McLAREN, D. J. 1977. The Silurian–Devonian Boundary Committee: A final Report. In MARTINSSON, A. (Ed.) *The Silurian–Devonian Boundary, IUGS Series A*, No. 5, Stuttgart, 1–34.

NEWELL, N. D. 1956. Fossil populations. In SYLVESTER-BRADLEY, P. C. (Ed.) The species concept in palaeontology. *Systematic Association Publication* **2**, 63–82.

PAUL, C. R. C. 1985. The adequacy of the Fossil Record reconsidered. *Special Papers in Palaeontology* **33**, 7–15.

PENNY, J. H. J. 1988. Early Cretaceous acolumellate pollen from Egypt. *Palaeontology* **31**, 373–418.

PHILIP, G. M. and WATSON, D. F. 1987. Some speculations on the randomness of Nature. *Mathematical Geology* **19**, 571–573.

PRANCE, G. T. and WHITE, F. 1988. The genera of Chrysobalanaceae: a study in practical and theoretical taxonomy, and its relevance to evolutionary biology. *Philosophical Transactions of the Royal Society, London* **320B**, 1–184.

PUBLICATIONS COMMITTEE OF THE PALAEONTOLOGICAL ASSOCIATION 1977. Notes for authors submitting papers. *Palaeontology* **20**, 921–929.

RIDE, W. D. L. *et al.* (Eds.) 1985. *International Code of Zoological Nomenclature*, 3rd edition. International Trust for Zoological Nomenclature, University of California Press, Berkeley and Los Angeles, 338 pp.

RIEDEL, W. R. and TWAY, L. E. 1989. Artificial intelligence programming for radiolarian research. In AGTERBERG, F. P. and BONHAM-CARTER, G. F. (Eds.) *Statistical Applications in the Earth Sciences*. Geological Survey of Canada, Paper 89–9.

RODGERS, J. 1959. The meaning of correlation. *American Journal of Science* **257**, 684–691.

SCOTT, G. H. 1985. Homotaxy and biostratigraphical theory. *Palaeontology* **28**, 777–782.

VOSS, E. G. *et al.* (Eds.) 1983. International Code of Botanical Nomenclature. *Regnum Vegetabile* **111**, 1–472.

ZHAMOIDA, A. I., KOVALEVSKY, O. P., MOISSEJEVA, A. I. and YARKIN, V. I. 1977. *Stratigraphic Code of the USSR*. Interdepartmental Stratigraphic Committee of the USSR, Leningrad, 148pp.

INDEX

Bold type is used to indicate the page number(s) of the main reference(s) to a subject